AS/A-LEVEL

Biology

Alan Clamp

ESSENTIAL WORD
DICTIONARY

To Philip and Avice Winfield, for their kindness and generosity.
To Elizabeth for the king-sized gin-and-tonics. But mostly to Helen,
who makes it all worthwhile.

Philip Allan Updates
Market Place
Deddington
Oxfordshire
OX15 0SE

Tel: 01869 338652
Fax: 01869 337590
e-mail: sales@philipallan.co.uk
www.philipallan.co.uk

ISBN 0 86003 372 4

Acknowledgements
I would like to thank David Cross for his faith in my abilities and Mary
Jones for her helpful advice on the manuscript. I would also like to thank
Roger Ransome, who didn't just teach biology — he taught his students
how to be biologists.

P00150
Printed by Raithby, Lawrence & Co Ltd, Leicester

Introduction

This is a concise dictionary of key terms used in AS and A-level biology. Each word has been carefully chosen and is explained in detail to enhance your knowledge and understanding of this subject.

Many other dictionaries will contain words that are not relevant to students of AS and A-level biology. This dictionary, however, only lists the most important terms and does not include words on the periphery of the subject. This will help you to learn the key terms more efficiently.

In the dictionary, each word is defined in up to four parts, as follows.
(1) A brief definition.
(2) Further explanation of the word.
(3) An example.
(4) An examiner's tip, such as where a word is commonly misunderstood, confused with another word, used in error, or found in conjunction with other words in the dictionary.

In many cases, all four parts are not needed and the entry has been amended accordingly. Finally, for each term it may be necessary to make a cross-reference to the words in italics in order to understand fully the entry you are reading.

Make extensive use of this 'essential words' dictionary. It could provide a significant boost to your chances of success in AS and A-level biology.

ABA: see *abscisic acid*.

abiotic factor: any of the non-living factors that make up the environment of living organisms.

■ *e.g.* Rainfall, soil *pH*, temperature and light intensity.

■ *TIP* This term is often used in conjunction with *biotic factor*, which refers to factors concerned with the living part of the environment.

abscisic acid (ABA): a plant growth substance which acts mainly as a growth inhibitor.

■ ABA stimulates the closing of *stomata* and may play a part in leaf fall in some deciduous trees. It also inhibits *germination* in *seeds*.

■ *TIP* Make sure that you know how ABA interacts with other plant growth substances, such as *auxins*, *cytokinins*, *ethene* and *gibberellins*.

absorption: the process by which dissolved substances are taken up by cells.

■ Absorption occurs across the cells lining the *ileum* and *nephrons* (in the kidney) and across root hairs from the soil. Solutes (dissolved substances) may move by simple *diffusion, facilitated diffusion, active transport* or *pinocytosis*.

■ *TIP* Many cells are adapted for efficient absorption by having a large surface area and by containing large numbers of *mitochondria*, which provide the *ATP* necessary for *active transport* and *pinocytosis*.

absorption spectrum: a graph showing the relative amounts of light of different wavelengths that are absorbed by a *pigment*.

■ *e.g.* The diagram below shows the *absorption spectrum* for *chlorophyll a*.

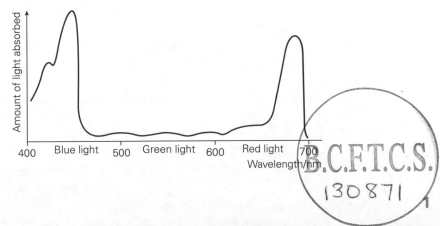

a

Chlorophyll absorbs light strongly at the blue and red ends of the spectrum, and reflects green light. This is why most leaves appear green.

■ *TIP* Students often confuse absorption spectrum with *action spectrum*, which shows the relative amounts of light of different wavelengths that are used in a particular process, such as *photosynthesis*.

accommodation: the process by which the eye focuses upon an object.

■ Accommodation occurs by reflex adjustments to the shape of the *lens* of the eye. This permits clear images of objects at a range of distances to be formed on the *retina*.

acetylcholine (ACh): a chemical that allows the transmission of an *impulse* from one *neurone* (nerve cell) to another.

■ Neurones connect with each other at a *synapse*. The arrival of a nerve impulse at the presynaptic membrane causes the release of ACh, which diffuses across the synaptic cleft to *receptors* on the postsynaptic membrane. This triggers another nerve impulse in the postsynaptic neurone.

acetylcholinesterase: an enzyme which causes the breakdown of *acetylcholine*.

■ After it has triggered a nerve impulse, acetylcholine must be removed by acetyl-cholinesterase. Otherwise, impulses would be continually generated in the postsynaptic cell.

acetylcoenzyme A: a chemical compound that is an important intermediate in *aerobic respiration*.

■ Acetylcoenzyme A links *glycolysis* and the *Krebs cycle* in aerobic respiration. *Pyruvate*, the end product of glycolysis, loses carbon dioxide to produce a two-carbon compound (acetyl). This is then linked to coenzyme A (temporarily forming acetylcoenzyme A), which feeds the two-carbon compound into the Krebs cycle.

ACh: see *acetylcholine*.

acid rain: this is produced when certain gases in the atmosphere, such as sulphur dioxide, dissolve in rain water to form acids.

■ Acid rain reduces the *pH* of soil, rivers and lakes, affecting *ecosystems*. Typical effects include the death of trees, particularly conifers, and aquatic organisms such as fish.

acrosome: a thin cap-like *vesicle* found at the tip of a *sperm*.

■ The acrosome contains *enzymes* which are released when the sperm meets an *ovum*. The digestive action of the enzymes allows the sperm to penetrate the membrane surrounding the ovum, leading to *fertilisation*.

actin: a protein that forms the thin filaments found in the microscopic fibres of skeletal muscle.

■ *TIP* This term is often seen in conjunction with *myosin*, a protein forming the thick filaments found in the *myofibrils* (microscopic fibres) of skeletal muscle.

action potential: the rapid change in electrical charge across the membrane of a nerve cell, causing the transmission of an *impulse*.

■ Stimulation of a nerve cell causes the electrical charge across the membrane to change from the *resting potential* of −60 mV to +40 mV. The change in electrical charge, which lasts for only a few milliseconds before the resting potential is restored, is called an action potential.

■ **TIP** Exam questions often ask how the movement of sodium and potassium ions across the nerve cell membrane result in the production of an action potential. It is important that you know this information, which is best summarised in the form of a flow-chart.

action spectrum: a graph showing the relative amounts of light of different wavelengths that are used in a particular process, such as *photosynthesis*.

■ The action spectrum for photosynthesis is shown in the diagram below. Note that it is similar to the *absorption spectrum* for *chlorophyll a*. This suggests that most of the light absorbed by chlorophyll is used in photosynthesis.

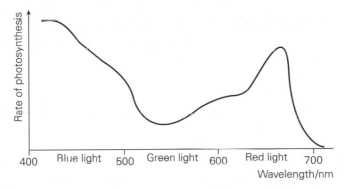

■ **TIP** Students often confuse action spectrum with absorption spectrum, which shows the relative amounts of light of different wavelengths that are absorbed by a *pigment*.

activation energy: the energy required to activate or begin a chemical reaction.

■ **TIP** You are most likely to come across this term when studying *enzymes* or *metabolism*. Enzymes speed up metabolic reactions because they bring the reactants together in such a way that the activation energy is lowered.

active site: the part of an *enzyme* to which a substrate becomes attached during a biochemical reaction.

■ Enzymes are globular proteins and the active site has a particular shape. Only a substrate with the corresponding shape will attach to the active site, hence the specificity of most enzymes. Changes in *pH*, or high temperatures, will denature the enzyme, altering the shape of the active site. If this happens, the substrate will no longer fit the site and the enzyme will cease to function.

■ **TIP** Do not write about enzymes being 'killed' by changes in pH or high temperatures. They are not living organisms, so they cannot die!

active transport: the movement of substances from where they are less concentrated to where they are more concentrated (against a concentration gradient).

■ Active transport is undertaken by carrier proteins in *cell membranes*, which move specific molecules or *ions* against the concentration gradient using energy supplied by *ATP*.

■ *e.g.* The reabsorption of *glucose* from the *proximal convoluted tubule* in the *kidney* is an example of active transport.

adaptation: any feature of the structure or physiology of an organism that makes it well suited to its *environment*.

■ Organisms that are better adapted to their environment are more likely to survive and reproduce than those that are less well adapted. If these adaptations are genetically controlled, the next generation will have a higher proportion of well-adapted individuals (assuming that there are no major environmental changes). Eventually, these adaptations will be shown by the vast majority of the *species*. *Natural selection* of inheritable adaptations may ultimately lead to the development of new species.

■ *e.g.* *Palisade cells* are adapted for efficient *photosynthesis*; marram grass (a *xerophyte*) is adapted to conserve water in arid environments; polar bears are adapted to live in cold climates.

adaptive radiation: the process by which a single ancestral type of organism gives rise to a number of different forms.

■ Adaptive radiation occurs during *evolution* when organisms diversify as they exploit different ecological *niches*. The process requires that a number of niches are available and that the different forms become *reproductively isolated* from one another.

■ *e.g.* Darwin's finches are a group of 14 species of finch found on the Galapagos Islands, about six hundred miles off the coast of Ecuador. Darwin believed that all the Galapagos finches were the descendants of a few finch-like birds that originally arrived from the mainland. Lack of competition and of predators allowed the finches to exploit many niches, feeding on seeds, insects, nectar and fruit, some nesting in trees and some on the ground.

■ *TIP* Adaptive radiation is also known as divergent evolution, as a number of species may diverge or evolve from the original ancestral species.

adenine: a *purine* nitrogenous base that is found in *DNA* and *RNA*.

■ In DNA, adenine always forms a bond with *thymine*, a *pyrimidine* nitrogenous base.

adenosine triphosphate: see *ATP*.

ADH: see *antidiuretic hormone*.

adrenal gland: one of a pair of glands found adjacent to the kidney which is responsible for secreting several important hormones.

■ *e.g.* The inner region of the gland (*medulla*) secretes *adrenaline*, following stimulation by the *sympathetic nervous system*; the outer region (cortex) secretes *aldosterone*, in response to adrenocorticotrophic hormone (ACTH) from the anterior *pituitary gland*.

adrenaline: a hormone secreted by the *adrenal glands* at times of stress.

■ Adrenaline has many physiological effects, aimed at preparing the body for action, such as increasing the heart rate and stimulating the conversion of *glycogen* to *glucose*.

adrenergic synapse: any *synapse* in which the *neurotransmitter* is *noradrenaline*.

■ Adrenergic synapses are found between motor neurones and *effectors* in the *sympathetic nervous system*.

■ *TIP* Try not to confuse this term with *cholinergic synapse*, which refers to synapses in which the neurotransmitter is *acetylcholine*.

aerobic respiration: the breakdown of *glucose* in the presence of oxygen to yield energy.

■ The process of aerobic respiration is summarised in the diagram below.

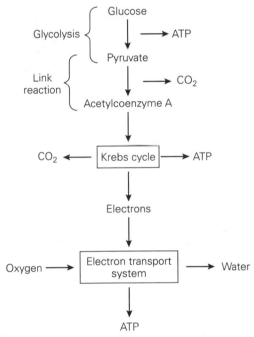

This can be summarised as:

glucose + oxygen + ADP + Pi ⟶ carbon dioxide + water + ATP

Respiration begins with *glycolysis*, which involves the breakdown of *glucose* to *pyruvate*, generating a small amount of *ATP*. If oxygen is present, pyruvate is converted to *acetylcoenzyme A* (releasing carbon dioxide) in the *link reaction*. Acetylcoenzyme A then enters the *Krebs cycle*, where ATP, carbon dioxide and reduced *NAD* are produced. The reduced NAD provides electrons that are passed along the *electron transport chain*, ultimately combining with oxygen to form water. The movement of electrons through the electron transport system releases a large amount of energy, which is used to produce further ATP.

■ *TIP* Make sure you know the difference between aerobic respiration, which requires oxygen, and *anaerobic respiration*, which takes place when oxygen is not available.

a

agonist: a chemical that binds to a *receptor* on a cell, triggering a physiological response.

■ Agonists are usually drugs, *hormones* or *neurotransmitters*.

■ *e.g.* Anatoxin is a poisonous chemical produced by algae, which is very similar in structure to *acetylcholine*. However, unlike acetylcholine, anatoxin is not broken down by *acetylcholinesterase*. Therefore, anatoxin stimulates continuous nerve *impulses* at a *synapse*, resulting in abnormal muscle function in animals consuming the poison.

■ *TIP* Try not to confuse agonistic chemicals, which mimic the effects of hormones or neurotransmitters, with *antagonistic* chemicals, which block these effects.

Agrobacterium: a type of *bacteria* commonly used in *gene technology*.

■ *Agrobacterium* is capable of entering plant cells and incorporating its *plasmids* into the *DNA* of the plant (often forming tumour-like growths known as crown galls). Genetic engineers can use these plasmids to introduce *genes* into crop plants, such as a gene for *herbicide* resistance. This means that the crop will be resistant to the herbicide and it can be sprayed safely on to crop fields to kill competing weeds.

aldosterone: a *hormone* involved in regulating the concentration of sodium *ions* in the blood.

■ Aldosterone is secreted by the *adrenal gland* in response to a fall in the sodium ion concentration of the blood. It increases the *active transport* of sodium ions from the kidney *nephrons* into the blood.

allele: one of the different forms of a *gene*.

■ *e.g.* The height of pea plants is controlled by a gene, which has two forms or alleles. One allele (T) is *dominant* and produces a tall plant; the other allele (t) is *recessive* and produces a short plant.

■ *TIP* Many students confuse the terms gene and allele. Remember that a gene controls a character, such as eye colour, and that alleles control different forms of this character, such as blue, green or brown eyes.

allopatric speciation: the development of one or more *species*, which occurs when populations of the parent species become geographically isolated from each other.

■ *TIP* A common exam question asks you to distinguish between allopatric and *sympatric speciation*, which occurs when populations living in the same area are prevented from interbreeding.

all-or-nothing: a term used to describe the fact that *action potentials* in nerve cells are always identical in size.

■ A stimulus will either trigger an action potential of a fixed size, or it will fail to trigger an *impulse* at all. A bigger stimulus will not trigger a bigger action potential. Therefore, the only way that information about the size of a stimulus can be transmitted is by changing the frequency of the impulses. A large stimulus will trigger high-frequency impulses and a small stimulus will trigger low-frequency impulses.

allosteric enzyme: an *enzyme* that exists in two interchangeable forms, one of which is active and the other inactive.

■ The binding of a (non-*substrate*) molecule at an allosteric site changes the shape of the *active site* of an enzyme. This may activate or, more commonly, inhibit the enzyme.

alternation of generations: a situation where the life cycle of an organism involves alternating *spore*-producing (*diploid*) and *gamete*-producing (*haploid*) generations.

■ The diagram below shows alternation of generations in the basic plant life cycle.

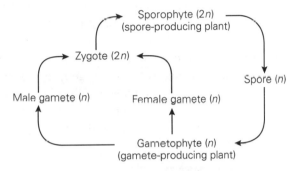

Alternation of generations is seen most clearly in *Bryophyta* (where the haploid [*n*] *gametophyte* is the dominant form) and *Filicinophyta* (where the diploid [*2n*] *sporophyte* is dominant). In flowering plants, the adult plant is a sporophyte and the gametophytes are simply the gamete-producing structures found within the *flowers*.

alveolus: one of the air sacs in the *lungs* where gas exchange takes place.

■ Oxygen diffuses from the alveolus into the *blood* and is transported to respiring cells. Carbon dioxide diffuses from the blood into the alveolus and is breathed out.

amino acid: the basic sub-unit or *monomer* from which *proteins* are formed.

■ The diagram below shows the general structure of an amino acid.

$$
\begin{array}{c}
H \\
\diagdown \\
N - C - C \diagup^{O} \\
\diagup \quad | \quad \diagdown \\
H \quad R\ group \quad OH
\end{array}
$$

Basic amino   Acidic carboxyl
group   group

The R group varies for each amino acid. There are twenty different amino acids that can be linked together by *condensation reactions* to form proteins. Certain amino acids must be supplied in the diet as they cannot be synthesised by the body. These are known as essential amino acids.

■ **e.g.** If R is a hydrogen atom (H), the amino acid is glycine; if R is a methyl group (CH_3), the amino acid is alanine.

a

amylase: an enzyme which breaks down *starch* into *maltose*.

◼ In the mammalian digestive system, amylase is secreted by the salivary glands and by the *pancreas*. It is also important in seeds, where it breaks down the stored starch during *germination*.

anabolism: the synthesis of complex molecules from simple molecules.

◼ Anabolism is really a series of chemical reactions (controlled by *enzymes*) requiring energy in the form of *ATP*.

◼ ***e.g.*** *Photosynthesis* and *protein synthesis*.

◼ ***TIP*** The opposite of anabolism is catabolism, the breakdown of complex molecules into simpler ones. Be careful not to confuse these two terms.

anaerobic respiration: the breakdown of *glucose* in the absence of oxygen to yield energy.

◼ Anaerobic respiration is basically *glycolysis* and only yields 5–6% of the *ATP* produced by *aerobic respiration* (which requires oxygen). In plants and *microorganisms* such as yeast, the *pyruvate* produced by glycolysis is converted to ethanol and carbon dioxide. In animals, the pyruvate is converted to lactic acid. The process of anaerobic respiration is summarised in the diagram below.

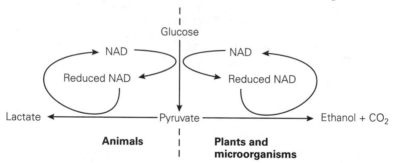

Angiospermophyta: the plant *phylum* which contains the flowering plants.

◼ Angiosperms share the following features:
  • the gametes are produced within *flowers*
  • *seeds* are formed as a result of *fertilisation* and are enclosed in *fruits*

◼ ***e.g.*** Orchids (*monocotyledons*) and oak trees (*dicotyledons*).

Animalia: the *kingdom* containing animals.

◼ Animals share the following features:
  • they are multicellular organisms
  • animal cells do not possess *cell walls*
  • they show *heterotrophic nutrition*

◼ ***e.g.*** Starfish (members of the *phylum Echinodermata*) and lizards (members of the phylum *Chordata*).

Annelida: the animal *phylum* containing earthworms and leeches.

◼ Annelids share the following features:
  • they possess a fluid-filled cavity called the *coelom* which acts as a hydrostatic skeleton for muscles to push against during locomotion

- they have long, cylindrical bodies which are divided up into distinct segments
- most segments possess small bristles (chaetae), which aid locomotion

antagonist: a chemical that binds to a *receptor* in a cell, blocking the normal physiological response.

■ Antagonists are usually drugs, *hormones* or *neurotransmitters*, which inhibit the effects of *agonists*.

■ **e.g.** The South American arrow poison, curare, blocks the action of *acetylcholine* at the junction between nerve and muscle. This leads to paralysis.

■ **TIP** Try not to confuse antagonistic chemicals, which block the effects of hormones or neurotransmitters, with agonistic chemicals, which mimic these effects. Antagonistic generally means 'opposite' and occurs elsewhere in biology, such as antagonistic muscles which have opposing actions.

anther: part of a *flower* in which *pollen* develops and is later released when the anther ruptures.

antibiotic: a substance produced by a living organism that kills or inhibits the growth of *microorganisms*.

■ Antibiotics work by affecting metabolic processes that occur only in micro-organisms such as *bacteria*. *Bactericidal* (biocidal) antibiotics kill bacteria, whereas *bacteriostatic* (biostatic) antibiotics simply prevent bacteria from reproducing.

■ **e.g.** Penicillin (prevents *cell wall* formation); streptomycin (inhibits *protein synthesis*); and ciprofloxacin (inhibits *DNA replication*).

■ **TIP** It can be easy to confuse the terms antibiotic, *antibody*, *antigen* and *antiseptic*. All four words are defined and explained in this dictionary.

antibody: a *protein* molecule produced by *lymphocytes* in response to stimulation by an *antigen*.

■ Antibodies are also known as immunoglobulins and exist in many different forms, each with a distinct chemical structure and biological role. An antibody can combine with a specific *antigen* to neutralise, inhibit or destroy the antigen.

■ **TIP** It can be easy to confuse the terms antibody, *antibiotic*, *antigen* and *antiseptic*. All four words are defined and explained in this dictionary.

anticodon: a triplet of bases in *transfer RNA (tRNA)* that can form base pairs with a specific *codon* during the synthesis of *proteins*.

■ During *translation*, tRNA transports an *amino acid* to a *ribosome* and the anticodon (e.g. UCG) attaches to the complementary codon (e.g. AGC), a triplet of bases in *messenger RNA (mRNA)*.

antidiuretic hormone (ADH): a *hormone* which makes the distal convoluted tubules and *collecting ducts* of a kidney *nephron* more permeable to water.

■ ADH is released by the *pituitary gland* in response to a fall in the *water potential* of the *blood*. Its effect on the *kidney* means that more water is reabsorbed into the blood (increasing its water potential) and less is lost in the *urine*. In other words, ADH prevents the production of large quantities of watery urine (diuresis), hence its name: antidiuretic hormone. The flow chart below summarises the way in which ADH helps to control water balance in the body.

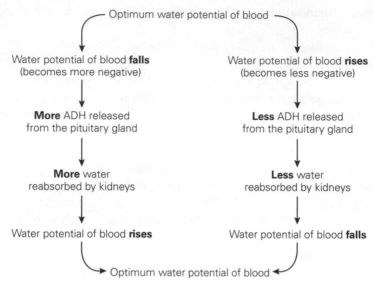

antigen: a molecule that triggers the production of *antibodies* from *lymphocytes*.

■ Antigens are usually proteins, although other types of molecule such as *polysaccharides*, *lipids* and *nucleic acids* can also act as antigens.

■ *TIP* It can be easy to confuse the terms antigen, *antibiotic*, *antibody* and *antiseptic*. All four words are defined and explained in this dictionary.

antiseptic: a substance that kills or inhibits the growth of *microorganisms* and can be used safely on the skin.

■ *e.g.* Hydrogen peroxide and ethanol.

■ *TIP* It can be easy to confuse the terms antiseptic, *antibiotic*, *antibody* and *antigen*. All four words are defined and explained in this dictionary. Antiseptics are not the same as disinfectants, which cannot be used on the skin.

aorta: the main *artery* of the body in mammals.

■ The aorta carries oxygenated blood from the left *ventricle* of the *heart* to all parts of the body except the *lungs*. All other arteries (except the *pulmonary artery*) branch off the aorta. The wall of the aorta contains baroreceptors which monitor blood pressure, and chemoreceptors which monitor changes in the concentration of oxygen and carbon dioxide in the blood.

apoplastic pathway: the route by which water and solutes travel through the *cell walls* of a plant.

■ The apoplastic pathway is the main route by which water moves through the roots to the *xylem*, and from the xylem into the leaves.

■ *TIP* Do not confuse the apoplastic pathway with the other two routes along which water can travel through plants: the *symplastic pathway*, through the cytoplasm of the cells; and the vacuolar pathway, through the cell *vacuoles*.

aqueous humour: the fluid found in the front part of the eye, between the *cornea* and the *lens*.

■ The aqueous humour is produced by the ciliary body. It supplies the cornea and lens with nutrients, and helps to maintain the shape of the eye.

arteriole: a vessel that receives blood from an *artery* and carries it to *capillaries*.

■ Arterioles have a smaller diameter than arteries, but have the same general structure. Many arterioles also have *sphincter muscles* where they join with the capillaries. This feature enables them to regulate the blood supply to particular capillary networks, according to the needs of the part of the body concerned.

artery: a vessel that carries blood away from the *heart*.

■ All arteries carry oxygenated blood, except the *pulmonary artery*, which transports blood from the heart to the *lungs*. The general structure of an artery is shown in the diagram below.

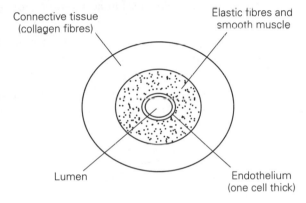

Arterial blood is under high pressure, so the walls of the arteries contain elastic tissue that helps to even out the pulsed flow. The walls of smaller arteries are particularly rich in *smooth muscle* fibres. This muscle may enable the arteries to alter their diameter and so helps to regulate the blood supply to a particular organ, although this happens mainly in *arterioles*.

Arthropoda: the animal *phylum* containing insects, spiders and crustaceans.

■ Arthropods share the following features:
  • a protective *exoskeleton* or cuticle made of *chitin* (this prevents growth, except by moulting, and limits the maximum size of arthropods)
  • jointed appendages which are used as limbs, mouthparts, sense organs, wings and reproductive organs

■ *e.g.* Ants, scorpions, crabs and millipedes.

asexual reproduction: reproduction involving the formation of new individuals from a single parent without the fusion of *gametes*.

■ *e.g.* Many organisms in the *kingdom Protoctista* (e.g. amoeba) reproduce by fission, the splitting of one organism into two separate individuals. Yeast cells reproduce asexually by budding. Mosses and ferns produce spores. In flowering plants, asexual reproduction is known as vegetative propagation, and can be seen in the development of bulbs in daffodils and tubers in potatoes.

■ *TIP* Make sure that you know the differences between asexual reproduction (which gives rise to genetically identical offspring) and *sexual reproduction* (which results in genetic variation among offspring).

atom: the smallest part of an *element* that cannot be broken down further by chemical means.

■ An atom is made up of a nucleus containing protons (particles with a positive charge) and neutrons (particles with no charge), surrounded by orbiting *electrons* (particles with a negative charge). Atoms may combine to form *molecules*.

■ *e.g.* A hydrogen atom (H) contains one proton and one electron. It can combine with another hydrogen atom and an oxygen atom (O), containing eight protons, eight neutrons and eight electrons, to form water (H_2O).

ATP (adenosine triphosphate): a molecule that acts as an energy carrier in all living cells.

■ ATP is a *nucleotide*, containing *ribose* as its pentose sugar and *adenine* as its nitrogenous base. Three phosphate groups are attached to the ribose sub-unit. The bond attaching the end phosphate group can be hydrolysed by ATPase to yield ADP (adenosine diphosphate) and a large quantity of energy. This reaction is reversible if enough energy is supplied, and is summarised in the diagram below.

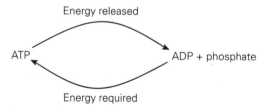

The energy required to produce ATP comes from *respiration* and *photosynthesis*. ATP has a number of roles in living organisms, including supplying the energy required for *protein synthesis, active transport* and contraction of *muscle*.

atrioventricular node (AVN): a specialised area of *muscle* between the *atria* and *ventricles* of the *heart*.

■ Electrical activity from the *sino-atrial node (SAN)* passes from the atria to the ventricles through the AVN. There is a short delay at the AVN that ensures that the atria have completely emptied their contents into the ventricles, before the ventricles contract. The AVN therefore plays an important role in coordinating the *cardiac cycle*.

■ *TIP* Try not to confuse the AVN with the SAN (pacemaker), which is found in the wall of the right atrium and initiates the heart beat.

atrioventricular valve: one of the valves between the *atria* and the *ventricles* of the heart.

■ The function of the valves is to ensure that blood flows in the correct direction through the heart. The valves are opened and closed by blood pressure at various points in the *cardiac cycle*, as shown in the table below.

Stage	Atrial pressure	Ventricular pressure	AV valves
Atrial systole	High	Low	Open
Ventricular systole	Low	High	Shut
Atrial and ventricular diastole	Low	Low	Open

The valve on the left side of the heart has two flaps and is called the *bicuspid valve*; the one on the right side has three flaps and is known as the *tricuspid valve*.

atrium: a chamber of the *heart* which receives blood returning from the organs of the body.

▨ In mammals, the left atrium receives oxygenated blood from the *lungs* via the *pulmonary vein*, and the right atrium receives deoxygenated blood from the other organs of the body via the *vena cava*.

▨ *TIP* Some textbooks may refer to the left and right auricle in mammals. An auricle is the same as an atrium. Do not confuse atrium with *ventricle*, which is a chamber of the heart that pumps blood to the organs of the body.

autonomic nervous system: the part of the nervous system that is responsible for controlling the involuntary activities of the body.

▨ The autonomic nervous system is divided into two sections:
- the *sympathetic nervous system* — responsible for preparing the body for action (e.g. increases heart rate)
- the *parasympathetic nervous system* — responsible for relaxing the body (e.g. decreases heart rate)

▨ *TIP* The sympathetic and parasympathetic systems tend to work antagonistically on the same organs. If you learn some of the functions of the sympathetic system, you will know that the parasympathetic system has the opposite effects.

autosome: a *chromosome* that is not a *sex chromosome*.

▨ In a human body cell there are 23 pairs of chromosomes. One pair of these are sex chromosomes and the other 22 pairs are autosomes.

autotrophic nutrition: the synthesis of organic molecules from simple inorganic molecules, such as carbon dioxide and water.

▨ Autotrophic nutrition requires a source of energy. *Photosynthesis* uses light energy (e.g. green plants) and *chemosynthesis* uses energy derived from chemical reactions (e.g. nitrifying *bacteria* in soil). Autotrophic organisms occupy the *producer trophic level* in *food chains* and *food webs*.

▨ *TIP* Do not confuse autotrophic nutrition with *heterotrophic nutrition*, in which an organism gains its required nutrients by consuming complex organic molecules.

a

auxin: a plant growth substance that acts mainly as a growth stimulator.

■ Auxins stimulate cell division, cell elongation, fruit development and (at low concentrations) root growth. They also inhibit lateral bud development, leaf and fruit abscission and (at high concentrations) root growth.

■ *TIP* Make sure that you know how auxins interact with other plant growth substances, such as *abscisic acid, cytokinin, ethene* and *gibberellin*.

AVN: see *atrioventricular node*.

axon: the elongated section of a *neurone* (nerve cell) which carries *impulses* away from the *cell* body.

bacteria: a group of microorganisms which play an important role in nutrient recycling, disease transmission and industrial processes.

■ Bacteria are classified as *prokaryotes*. They are small in size (a mean diameter of less than 5 μm), do not have a *nucleus* or any other membrane-bound *organelles*, and their *DNA* is in circular strands in the *cytoplasm*. A diagram of a typical bacterial cell is shown below.

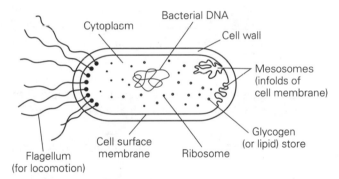

Bacteria mainly reproduce by cell division, although a few also use a type of sexual reproduction called conjugation.

■ *e.g.* *Vibrio cholerae* causes the disease cholera. *Streptococcus lactus* is used in making cheese. *Bacillus subtilis* is used in genetic engineering.

bactericidal: describes a chemical substance that kills *bacteria*.

■ *e.g.* Disinfectants and *antiseptics*.

■ *TIP* It is easy to confuse this term with *bacteriostatic*.

bacteriostatic: describes a chemical substance that prevents the reproduction of *bacteria*.

■ *e.g.* Most *antibiotics* are bacteriostatic.

■ *TIP* It is easy to confuse this term with *bactericidal*.

balanced diet: a diet that includes all the necessary nutrients in the required amounts to maintain good health.

■ In humans, a balanced diet contains *proteins*, *carbohydrates*, *lipids*, *vitamins*, minerals, water and dietary fibre. The required amounts of these nutrients

b

depend partly on the age, sex (including pregnancy and *lactation*) and level of activity of the individual concerned.

basal metabolic rate (BMR): the amount of energy required to sustain an animal at rest.

■ BMR is a measure of the energy required for vital activities that keep the body alive, such as the action of the heart and the muscles associated with breathing. Values for BMR are usually given as energy units per square metre of body surface per hour ($kJ\,m^{-2}h^{-1}$). This takes into account variation in body size. There are four main factors that affect BMR:

- age — BMR falls more-or-less continuously throughout life
- sex — females have a slightly lower BMR than males
- body size — small organisms tend to have a higher BMR per unit weight than large ones
- *thermoregulation* — *endotherms* ('warm-blooded' animals) usually have a higher BMR than *ectotherms* ('cold-blooded' animals)

B-cell: a *lymphocyte* which produces *antibodies* when stimulated by an *antigen*.

■ Each B-cell has receptors on its surface which will only bind to a specific antigen. In the presence of this antigen, the B-cell divides to form a large number of *plasma* cells. These cells produce antibodies which destroy the antigen. B-cells are therefore important in the *immune* response of an organism.

Benedict's test: a *biochemical test* that can be used to show the presence of a reducing sugar.

■ The test involves heating a solution of the unknown sample with Benedict's reagent. An orange–red precipitate indicates that a reducing sugar (e.g. *glucose*) is present.

bicuspid valve: the valve between the left *atrium* and the left *ventricle* in the heart of a mammal.

■ *TIP* Do not confuse the bicuspid valve with the *tricuspid valve*, which is found between the right atrium and the right ventricle. It may help to think about the t**R**icuspid valve being on the **R**ight. Both valves are also known as atrio-ventricular valves.

bilateral symmetry: a bilaterally symmetrical organism is one that can be divided in only one plane to give two approximately identical halves.

■ Most animals are bilaterally symmetrical and can be divided up into left and right halves. Members of the phyla *Cnidaria* and *Echinodermata* are exceptions to this rule.

■ *TIP* Note the difference between this term and *radial symmetry*, which describes organisms that can be divided in any longitudinal plane to give two identical halves.

bile: an alkaline solution that is released into the small intestine to aid the *digestion* of fat.

■ Bile is produced in the *liver* and stored in the *gall bladder* between meals. Its release is stimulated by the hormone cholecystokinin, when partly digested food from the stomach enters the small intestine. Bile emulsifies fats, which means that it breaks down fat into small droplets, increasing the surface area for the action of *lipase* enzymes. It also neutralises stomach acid and provides the optimum *pH* for pancreatic enzymes (pH 7–8).

■ *TIP* Remember that there are no digestive *enzymes* in bile.

bioaccumulation: the process by which certain chemicals become concentrated in the bodies of animals found at the top of food chains.

■ *e.g.* In the early 1950s, the chemical DDT was widely used to control insect pests. Insects containing DDT were eaten by small birds, which did not break the chemical down, but stored it in body fat. Birds of prey, such as sparrow-hawks, then consumed relatively large numbers of the smaller birds. This process resulted in further bioaccumulation of DDT in the tissues of the birds of prey, eventually building up to lethal concentrations.

biochemical oxygen demand (BOD): the quantity of oxygen that is removed by microorganisms from a sample of water in a given time.

■ BOD is a useful measure of aquatic *pollution*. Water containing a large quantity of organic matter, such as sewage, also contains large numbers of *bacteria*. The metabolic activities of the bacteria remove oxygen from the water, resulting in a high BOD.

biochemical tests: a group of simple tests that can be used to identify important biological molecules.

■ The details of the biochemical tests are summarised in the table below.

Test	Procedure
Biuret test for proteins	Add sodium hydroxide to the test sample, followed by a few drops of copper sulphate solution. If protein is present the solution turns purple.
Emulsion test for lipids	Shake the test sample with ethanol and then pour the resulting solution into a test tube containing water. If lipid is present a white emulsion is formed.
Benedict's test for reducing sugars (e.g. glucose)	Heat the test sample with Benedict's reagent. If a reducing sugar is present an orange–red precipitate is formed.
The test for non-reducing sugars (e.g. sucrose)	First heat the test sample with Benedict's reagent to confirm that there is no reducing sugar present. Then hydrolyse the sugar by heating with dilute acid followed by neutralising with sodium hydrogen carbonate. Finally, repeat the Benedict's test. If a non-reducing sugar is present, the second test will give an orange–red precipitate.
Iodine test for starch	Add iodine solution to the test sample. If starch is present the solution turns blue–black.
Schultze's test for cellulose	Add Schultze's solution to the test sample. If cellulose is present the solution turns purple.

■ *TIP* These tests may be referred to by the test name, such as the biuret test, or simply as 'food tests'.

b

biological control: the use of a parasite or a predator to control the number of pest organisms in a particular area.

■ Biological control also includes breeding pest-resistant crops. It has a number of advantages over chemical control:
- it is very specific
- it needs fewer applications and is therefore often cheaper than chemical control
- pests do not usually become resistant to biological control
- it does not pollute the environment

However, there are also disadvantages to biological control:
- it is relatively slow compared to chemical control
- pests are never completely eliminated, so there is always some damage to crops
- the control organism may become a pest in its own right

■ *e.g.* Aphids (insect pests of plants) are controlled by the use of ladybird beetles.

■ *TIP* Comparing the advantages of chemical control, using *pesticides* and biological control, is a common examination question.

biomass: the mass of all the organisms present in a given area.

■ Biomass is usually expressed as mass per unit area, such as $g\,m^{-2}$. (See also *pyramid of numbers/biomass/energy*.)

■ *TIP* It is important to distinguish fresh (live) biomass and dry biomass, where water has been removed from the sample. The amount of water in living organisms is very variable, so dry biomass is usually a more accurate measure of the mass of living material. However, fresh biomass is often more convenient, especially if it is important to keep the organisms alive.

biotechnology: the use of microorganisms or biochemical reactions to make useful products or to carry out industrial processes.

■ *e.g. Enzymes* are used in biological washing powders.
Yeast is used in the production of bread and wine.
Gene technology is used to modify *bacteria* to synthesise human *insulin*.

biotic factor: any environmental factor that is associated with living organisms.

■ *e.g.* Predation, competition and food availability.

■ *TIP* This term is often used in conjunction with *abiotic*, which refers to factors concerned with the non-living part of the environment.

biuret test: a *biochemical test* that can be used to show the presence of a protein.

■ The test involves adding sodium hydroxide to the unknown sample, followed by a few drops of copper sulphate solution. A purple colour indicates that protein is present.

■ *TIP* You do not need to heat the mixture in this biochemical test.

blood: a suspension of cells in solution that acts as a transport medium within an animal.

■ In humans, blood consists of about 55% by volume of *plasma* and 45% by volume of cells. The cells consist of *red blood cells* (erythrocytes), *white blood*

b

cells (leucocytes) and small cell-fragments called *platelets*. Blood has a number of important biological functions, including the following:

- protection against disease — white blood cells produce *antibodies* and remove microorganisms by *phagocytosis*; platelets are important in *blood clotting* to prevent the entry of microorganisms
- transport — of nutrients, such as *glucose* and *amino acids*; of respiratory gases (oxygen and carbon dioxide); of excretory products, such as *urea*; of hormones, such as *insulin*; and of heat

blood circulation: the system that transports *blood* around the body of many animals.

■ All blood circulatory systems have a pump (the *heart*) and most have a series of blood vessels (*arteries, veins* and *capillaries*). The blood system in fish is known as a *single circulation* as blood passes through the heart only once in its passage around the body. Mammals, on the other hand, have a *double circulation* as blood passes through the heart twice as it travels around the body.

blood clotting: the production of a blood clot following damage to body tissues.

■ When tissues are damaged, soluble fibrinogen in blood *plasma* is converted into insoluble *fibrin* by a mechanism summarised in the diagram below.

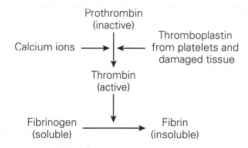

Fibrin forms a mesh over the surface of a wound that stops bleeding by forming a clot. The clot also helps to prevent the entry of microorganisms into the body.

blood glucose level: the concentration of glucose circulating in the blood plasma.

■ The concentration of glucose in the blood is typically 3.4–5.6 millimoles per dm^3. It is important to maintain this concentration between certain limits and the hormones *insulin* and *glucagon* play an important role in this process. If the glucose concentration falls too low (*hypoglycaemia*), the *central nervous system* ceases to function correctly. If it rises too high (*hyperglycaemia*), there will be a loss of glucose from the blood in the *urine*.

■ *TIP* The regulation of glucose concentration in the blood is a good example of *homeostasis*.

blood groups: the many types into which an individual's blood may be classified, based on the presence or absence of certain *antigens* on the surface of *red blood cells*.

■ In humans, the most important blood group system is the ABO system. There are four blood groups in this system, A, B, AB and O. Individuals with blood group A have the A antigen on their red blood cell membranes; those with blood group B have antigen B; AB individuals have both antigen A and antigen B; and group O has neither of these antigens. Humans also have *antibodies* present in their plasma. For example, an individual with blood group A will also have antibody b. If this person was given a transfusion of blood group B, the antigen B and antibody b would combine, causing the red blood cells to stick together (agglutinate). This could lead to blockage of blood vessels and possibly death. The table below summarises the antigens and antibodies present in the blood groups of the ABO system, together with the transfusions that each group could safely receive.

Blood group	Antigens	Antibodies	Safe transfusion groups
A	A	b	A or O
B	B	a	B or O
AB	A and B	none	A, B, AB or O
O	none	a and b	O

Note that because group AB can safely receive transfusions from any group, it is known as the universal recipient. Likewise, group O is known as the universal donor, because it can be transfused safely into any recipient.

BMR: see *basal metabolic rate*.

BOD: see *biochemical oxygen demand*.

Bohr effect: the effect of carbon dioxide concentration on the release of oxygen from *haemoglobin* in *red blood cells*.

■ The affinity of haemoglobin for oxygen is represented by the *oxygen dissociation curve*. An increase in the concentration of carbon dioxide (or a decrease in *pH*) causes the curve to shift to the right, causing haemoglobin to release more oxygen. Likewise, a decrease in carbon dioxide concentration (or an increase in pH) makes it less likely that haemoglobin will release oxygen. Therefore, in actively respiring tissues, where the concentration of carbon dioxide in the blood is high, haemoglobin readily releases its oxygen. In the lungs, on the other hand, where the concentration of carbon dioxide in the blood is low, haemoglobin readily binds to oxygen.

bone: a supporting tissue found in the skeletons of most vertebrates.

■ Bone is made up of cells embedded in a matrix of *collagen* fibres impregnated with calcium phosphate and other salts. The arrangement of the materials within the bone matrix gives it great strength and enables it to support the organism as well as protecting vital organs.

Bowman's capsule: the cup-shaped end of a kidney *nephron*.

■ Bowman's capsule (also known as the renal capsule) collects filtrate from the *glomerulus* in the kidney and therefore plays an important role in *ultrafiltration*.

brain: the organ which is responsible for controlling bodily functions by coordinating the activities of the nervous system.

■ The structure and functions of the major parts of the human brain are shown in the diagram below.

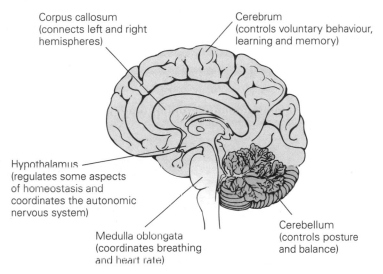

bronchiole: a small tube in the lungs that carries air between a *bronchus* and a large number of *alveoli*.

■ Apart from their smaller size, bronchioles differ from bronchi in that they do not have *cartilage* in their walls.

bronchus: a large tube in the lungs that carries air between the *trachea* and a number of *bronchioles*.

■ The walls of the bronchi, like those of the trachea, contain rings of *cartilage* for support and are lined with ciliated *epithelial cells*. Particles in the air that is inhaled are trapped in *mucus* on the surface of these cells and moved up to the throat by the action of the *cilia*. The mucus is then swallowed (and digested), or coughed out. This protective mechanism helps to prevent dust particles and microorganisms from entering the delicate parts of the lungs.

Bryophyta: the plant *phylum* which contains mosses and liverworts.

■ Members of this phylum share the following features:
- they have no true roots or leaves
- they require a moist *habitat* for survival and reproduction
- they have no vascular tissue
- they show alternation of generations, with the spore-bearing form (*sporophyte*) being dependent upon the dominant gamete-bearing form (*gametophyte*)

buffer: a solution or substance which resists changes in *pH* by taking in or releasing hydrogen ions.

■ The pH of a buffer solution changes very little following the addition of small amounts of either an acid or an alkali. If an acid is added, the excess hydrogen

b

ions are taken up by the buffer. If an alkali is added, hydrogen ions are released by the buffer to neutralise excess hydroxyl (OH⁻) ions.

■ *e.g.* *Haemoglobin*, *plasma proteins*, hydrogencarbonates and phosphates are all important buffers which help to control the pH of *blood* and *tissue fluid*. Buffer solutions are also used in experiments where it is necessary to maintain a particular pH value.

bundle of His: specialised cardiac muscle fibres that run from the *atrioventricular node* to the base of the heart.

■ During the *cardiac cycle*, electrical activity passes from the *sino-atrial node*, through the atrioventricular node, down the bundle of His and up through the walls of the ventricles in *Purkyne tissue*. This pattern of electrical activity causes the heart to contract in a coordinated manner, squeezing blood up from the bottom of the heart into the *aorta* and *pulmonary artery*.

C₃ plant: a plant in which the first product of *photosynthesis* is a three-carbon compound.

■ In the *light-independent* stage of photosynthesis, carbon dioxide is accepted by the five-carbon molecule, *ribulose bisphosphate*. The product of this reaction then immediately splits into two molecules of glycerate 3-phosphate (a three-carbon compound).

■ *e.g.* Most temperate plants, such as wheat and sugar beet.

C₄ plant: a plant in which the first product of *photosynthesis* is a four-carbon compound.

■ In the *light-independent* stage of photosynthesis, carbon dioxide is accepted by the three-carbon molecule, phosphoenolpyruvate (PEP). This forms the four-carbon compound malate. C_4 plants are adapted for tropical regions, where the concentration of carbon dioxide during the day is often a factor limiting the rate of photosynthesis. Malate is mainly produced at night, when there is more carbon dioxide available, and it acts as a store of carbon dioxide. When carbon dioxide is needed during the day, malate is decarboxylated to recycle PEP and release CO_2 for use in the *Calvin cycle*. The C_4 pathway also allows plants to conserve more water and avoids the problem of photorespiration, where oxygen inhibits the fixation of CO_2.

■ *e.g.* Many tropical plants, such as maize and sugar cane.

calcium: an essential mineral nutrient for both animals and plants.

■ Calcium has a number of functions:

● its compounds form structural components of the *cell walls* of plants (calcium pectate) and the *bones* and teeth of mammals (calcium phosphate)

● it plays an important role in *blood clotting*, muscle contraction and transmission across *synapses* in the nervous system

● it functions as an *enzyme* activator in many metabolic processes

Calvin cycle: a series of biochemical reactions, forming part of the *light-independent* stage of *photosynthesis*, which takes place in the stroma of the *chloroplasts* of C₃ *plants*.

■ The Calvin cycle involves the fixation of carbon dioxide and its subsequent reduction to carbohydrate.

The main steps in the Calvin cycle are summarised in the diagram below.

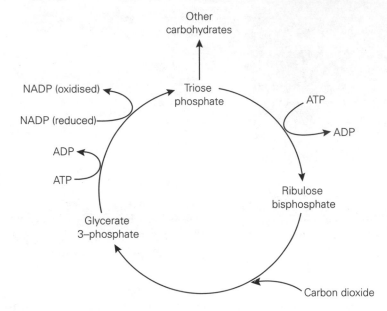

cAMP: see *cyclic AMP*.

capillary: a very small blood vessel where water, solutes and respiratory gases are exchanged with body tissues.

■ Capillaries carry blood from *arterioles* to *venules*. Their walls are only one cell thick, so that oxygen and nutrients can pass easily into the surrounding tissues. Similarly, waste materials, such as urea and carbon dioxide, can enter capillaries from the tissues and be transported away for excretion.

capsid: the outer protein coat of a *virus*.

■ The capsid is made up of units called capsomeres. Its main function is to protect the *nucleic acid* inside the virus. In some viruses, the capsid also contains specific receptor molecules which enable the virus to recognise and attach to host cells.

capsule: a thick layer surrounding the *cell wall* of certain bacteria.

■ Capsules are usually made of protein or *polysaccharides*. They appear to protect the bacteria by resisting *phagocytosis* and preventing desiccation. In general, therefore, bacteria with capsules are more likely to be pathogenic (disease-causing).

carbohydrate: a compound containing the elements carbon, hydrogen and oxygen, with the general formula $C_x(H_2O)_y$.

■ The simplest carbohydrates are *monosaccharides*, which usually contain three carbon atoms (trioses), five carbon atoms (pentoses) or six carbon atoms (hexoses). *Disaccharides* are formed from a *condensation* reaction between two hexose monosaccharides; and *polysaccharides* are made up of many hexose monosaccharides linked by condensation reactions. The following table summarises a number of biologically important carbohydrates.

C

Carbohydrate	Examples	Monosaccharide sub-units	Role in living organisms
Monosaccharides	Triose	N/A	Important intermediate in metabolism
	Glucose	N/A	Energy source in respiration
	Ribose	N/A	Part of RNA
Disaccharides	Sucrose	Glucose + Fructose	Transported in the phloem of plants
	Lactose	Glucose + Galactose	Main carbohydrate in milk
	Maltose	Glucose	Found in germinating seeds
Polysaccharides	Glycogen	α-Glucose	Storage carbohydrate in animals
	Starch	α-Glucose	Storage carbohydrate in plants
	Cellulose	β-Glucose	Part of plant cell walls

carbon cycle: one of several nutrient cycles, describing how the element carbon cycles in the environment.

■ The carbon cycle is summarised in the diagram below.

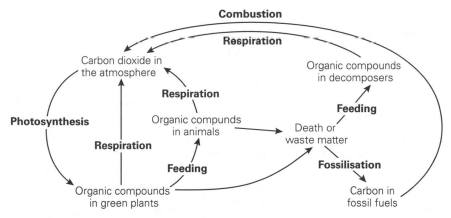

carbonic anhydrase: an enzyme that catalyses the reaction between carbon dioxide and water to form carbonic acid.

■ Once formed, carbonic acid will dissociate to form hydrogen ions and hydrogen-carbonate ions. This can be summarised by the equation below.

$$CO_2 + H_2O \xrightarrow{\text{carbonic anhydrase}} H_2CO_3 \longrightarrow H^+ + HCO_3^-$$

carbon dioxide   water   carbonic acid   hydrogen ion   hydrogencarbonate ion

Carbon dioxide produced in respiring tissues diffuses into *red blood cells*. Carbonic anhydrase in these cells converts the carbon dioxide into hydrogencarbonate ions, which diffuse into the plasma and are transported to the lungs. Here, the reverse series of reactions occur and carbon dioxide is expired.

cardiac cycle: the sequence of events taking place in a single heartbeat.

■ The heartbeat is initiated by a small area of muscle tissue in the wall of the right *atrium* known as the *sino-atrial node (SAN)* or pacemaker. This causes the following events:

- atrial systole — the atria contract, forcing blood into the *ventricles*
- ventricular systole — the ventricles contract, forcing blood out of the *heart*
- atrial and ventricular diastole — the atria and ventricles relax and the heart refills with blood

Atrial and ventricular diastole is then followed by atrial systole, and the cycle continues. The speed of the cycle (controlled by the SAN) can be influenced by two nerves from the brain: the cardiac accelerator nerve (part of the *sympathetic nervous system*) increases the rate at which the heart beats; and the vagus nerve (part of the *parasympathetic nervous system*) decreases the heart rate.

cardiac muscle: a specialised type of muscle, found only in the *heart* of vertebrates.

■ The microscopic structure of cardiac muscle is similar to that of *skeletal muscle*. However, there are numerous cross-links between the fibres and the muscle is myogenic, which means that it contracts and relaxes without external means of stimulation.

cardiovascular centre: part of the medulla oblongata in the *brain* which is responsible for regulating the heart rate.

■ The cardiovascular centre is made up of the cardiac acceleratory centre, which is responsible for increasing the heart rate, and the cardiac inhibitory centre, which slows the heart down.

carotid body: a collection of cells found in the walls of the carotid *arteries* that are sensitive to the oxygen concentration and pH of the *blood*.

■ A fall in the pH of the blood (caused by an increase in the concentration of carbon dioxide), or a fall in the oxygen concentration of the blood, causes the carotid bodies to send signals to the *brain* to increase the rate of breathing. Similarly, an increase in blood pH, or a rise in oxygen concentration, causes a decrease in the rate of breathing.

carrier molecule: a molecule that either transports substances across a membrane or carries electrons through an *electron transport chain*.

■ Carriers that transport molecules into or out of a cell are usually proteins, and they play an important role in *facilitated diffusion* and *active transport*. Electron carriers are also usually proteins, bound to a non-protein group, such as flavoproteins and *cytochromes*.

carrying capacity: the maximum *population* that a particular habitat can support.

■ Populations tend to grow until they reach the carrying capacity, at which point the birth rate is equal to the death rate. If the carrying capacity is exceeded, overcrowding and competition will eventually reduce the population.

cartilage: a hard-wearing and flexible connective tissue which plays an important role in the joints of the skeleton.

■ Cartilage, like *bone*, consists of cells embedded in a matrix of *collagen* fibres. However, unlike bone, the matrix is not impregnated with salts and does not contain any blood vessels. Cartilage is more compressible than bone and is therefore mainly used as a shock-absorber at the ends of bones in *synovial joints* and between the vertebrae.

Casparian strip: a band of impermeable *suberin* found in the walls of endodermal cells in plant roots.

▦ The *endodermis* is a ring of cells between the outer part of the root and the *vascular* tissue at the centre of the root. The Casparian strip prevents water and solutes from entering the *xylem* via the *apoplastic pathway* (through the cell walls and intercellular spaces). As a result, these substances have to pass along the *symplastic pathway* (through the cytoplasm of the endodermal cells). This allows the plant to control the movement of water and solutes into the xylem.

cell: the basic structural and functional unit of most living organisms.

▦ A cell comprises a mass of jelly-like *cytoplasm* which contains a number of *organelles* and is surrounded by a *cell membrane*. There are two basic types of cell:
- *prokaryotic* cells have no *nucleus* and very few *organelles* (e.g. *bacteria*)
- *eukaryotic* cells have a nucleus and a large number of different organelles (e.g. human liver cells and palisade mesophyll cells in the leaves of green plants)

The body of an individual organism contains many different types of cell, each type being specialised and adapted for a particular function. A group of similar cells which carry out a given function is known as a tissue (e.g. *muscle* tissue). Different tissues make up organs (e.g. the *stomach*) and several organs can combine to form a system (e.g. the digestive system).

▦ *TIP* A common examination question asks candidates to state similarities and differences between prokaryotic and eukaryotic cells.

cell cycle: the sequence of events which occur during cell growth and cell division.

▦ The cell cycle can be divided into four main stages, as shown in the figure below.

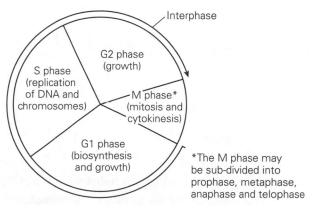

cell fractionation: the separation of the different components of a cell.

▦ Cell fractionation involves breaking up cells and separating the different *organelles* by ultracentrifugation. This involves spinning the cell contents at high speed. The larger organelles, such as nuclei, will tend to gather at the bottom of a *centrifuge* tube. Smaller organelles, such as *mitochondria*, will remain in suspension. The smaller organelles can then be separated by spinning the suspension

C

at higher speeds. Using this technique, it is possible to get a suspension containing a single type of organelle.

cell membrane: a membrane which encloses a *cell* or which surrounds certain *organelles* within a cell.

■ The membrane surrounding the cell is known as the cell surface membrane (or plasma membrane). Organelles such as the *nucleus, mitochondria* and *chloroplasts* have two membranes that form an envelope, others have just one membrane (e.g. *lysosomes*) and some have none (e.g. *ribosomes*). Cell membranes are typically about 7 nm thick and mainly consist of *phospholipids* and *proteins,* arranged as a *fluid mosaic.* The membranes are partially permeable, regulating the movement of substances in and out of the cell or organelles. Cell membranes also function as the site of metabolic reactions, such as the *electron transport chain* on the inner membranes of *mitochondria.*

cell sap: the solution (of sugars, amino acids and mineral salts) contained in the *vacuoles* of plant cells.

■ Cell sap plays an important part in maintaining the cell in a turgid state and providing the plant with support.

cellulose: a *polysaccharide* comprising long, unbranched chains of *glucose* molecules.

■ The structure of cellulose is shown in the diagram below.

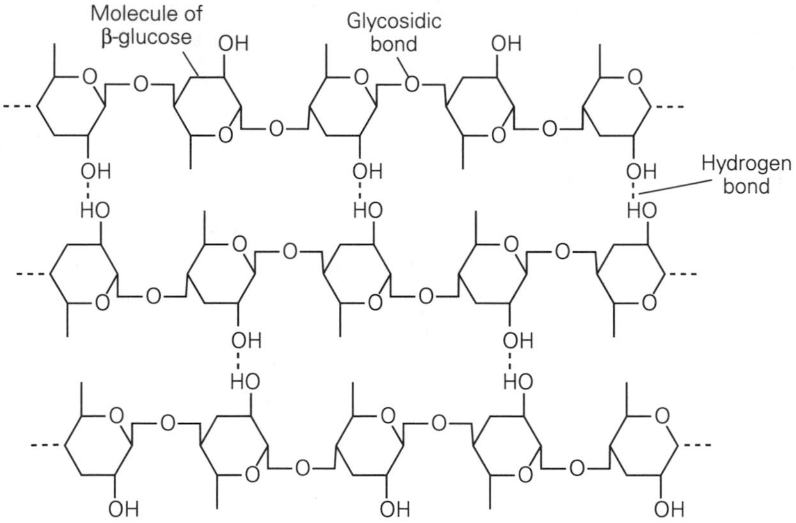

Cellulose chains are arranged in bundles called microfibrils. These microfibrils are cemented together in a matrix of other substances, forming strong but permeable *cell walls* in plants.

cell wall: a rigid structure found outside the cell surface membrane of plant, fungal and bacterial cells.

■ The main component of cell walls in plants is *cellulose*; in *bacteria* it is peptido-glycan (a complex polymer of *polysaccharides* and *amino acids*); and in fungi it is often the nitrogen-containing polysaccharide chitin. Many plant cell walls are further strengthened by the addition of *lignin*. Cell walls are very strong and provide support for the cell and for the whole organism. Nevertheless, they are also freely permeable to water and many other substances.

■ *TIP* A cell wall is not the same as a cell membrane.

central nervous system (CNS): the CNS consists of the *brain* and *spinal cord*, and is responsible for coordinating the activities of the nervous system.

centrifuge: a machine that separates materials of different densities by spinning them in a tube at high speed.

■ Centrifugation results in denser materials being forced to the bottom of the tube, while less dense substances remain nearer to the top of the tube. This process can be used for *cell fractionation* or to separate the major components of blood.

centrioles: structures found in animal cells which play a role in spindle-formation during cell division.

■ Centrioles are usually found in pairs. They are small, hollow cylinders, each containing a ring of microtubules.

centromere: the part of a *chromosome* that attaches to the spindle during cell division.

■ The centromere is responsible for holding daughter *chromatids* together during the early stages of cell division.

CFCs: see *chlorofluorocarbons*.

chemosynthesis: the synthesis of complex organic molecules from simple inorganic molecules, using the energy from chemical reactions.

■ *e.g.* ***Nitrobacter*** is a nitrifying bacterium found in soil which converts nitrites into nitrates as part of the *nitrogen cycle*. This chemical reaction releases energy, which enables the bacteria to produce *carbohydrates* and other organic compounds from carbon dioxide and water.

chiasma(ta): the point(s) at which the *chromatids* of *homologous chromosomes* cross over each other during *meiosis*.

■ Chiasmata occur during prophase I of meiosis. The chromatids break and re-join in a process called *crossing-over*. This results in the exchange of genetic material between the chromatids and new combinations of *genes* in the *gametes* formed during meiosis. Chiasmata are very important as they produce genetic *variation*.

chitin: a nitrogen-containing *polysaccharide*.

■ Chitin has a structural function in the exoskeleton of most insects and in the *cell walls* of some *fungi*.

chlorofluorocarbons (CFCs): manufactured chemicals that act as atmospheric pollutants, degrading the ozone layer.

■ CFCs were developed for use as propellants in aerosols and as coolants in refrigerators. Depletion of the ozone layer in the stratosphere increases the amount of ultraviolet light reaching the earth's surface and so increases the risk of skin cancer.

chlorophyll: a green pigment found in photosynthetic organisms.

■ Chlorophyll is actually a group of pigments, each consisting of a complex organic molecule containing magnesium. The most common forms are chlorophyll a (a blue–green pigment) and chlorophyll b (a yellow–green pigment). Chlorophyll molecules are responsible for the capture of light energy (see *absorption spectrum*) and therefore play a key role in the *light-dependent* stage of *photosynthesis*.

chloroplast: an *organelle* found in the cells of plants and algae which is the site of *photosynthesis*.

■ The diagram below shows the structure of a chloroplast.

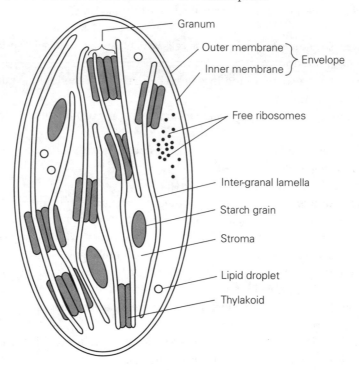

Granum

Outer membrane ⎤
Inner membrane ⎦ Envelope

Free ribosomes

Inter-granal lamella

Starch grain

Stroma

Lipid droplet

Thylakoid

cholesterol: a steroid which performs a number of functions in living organisms.

■ Cholesterol is an important component of *cell membranes* and is used to manufacture *bile* and steroid *hormones*, such as *testosterone* and *progesterone*. Cholesterol is produced in the liver and also consumed in the diet. High levels of cholesterol in the blood are associated with *coronary heart disease*.

cholinergic synapse: any *synapse* in which the *neurotransmitter* is *acetylcholine*.

■ Cholinergic synapses are found at *neuromuscular junctions* and in the *parasympathetic nervous system*.

■ *TIP* Try not to confuse this term with *adrenergic*, which refers to synapses in which the neurotransmitter is *noradrenaline*.

Chordata: the animal *phylum* containing fish, amphibians, reptiles, birds and mammals.

C

■ Chordata share the following features:
- a rod-like structure made of *cartilage*, called a notochord
- gill slits (visceral clefts) at some stage in their lifecycle
- a post-anal tail
- a dorsal hollow nerve cord

chromatid: one of the two identical strands of genetic material that make up a *chromosome*.

■ During the early stages of cell division, chromosomes consist of a pair of chromatids held together by a *centromere*. Chromosomes are not visible in non-dividing cells, when they assume an elongated form of *DNA* wrapped around proteins, known as chromatin.

■ *TIP* Try not to confuse the terms chromatid, chromatin and chromosome.

chromatography: a technique used to separate the individual components of a mixture.

■ Chromatography separates substances on the basis of their different solubilities in a solvent. The simplest form is paper chromatography of a solution of the mixture, such as a mixture of photosynthetic pigments extracted from leaves. The solution is applied at a point just above the bottom of a strip of chromatography paper, which is then suspended so that its end just dips into a suitable solvent. The solvent moves up the paper, separating the components of the mixture based on their different solubilities. It is then possible to identify the individual components in the mixture. The process of chromatography is summarised in the diagram below.

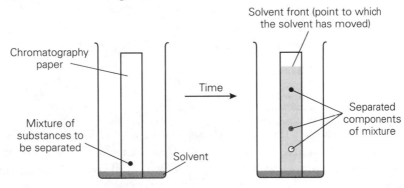

chromosome: a thread-like structure, composed of *DNA* and protein, found in the *nucleus* of plant and animal cells.

■ Each chromosome carries a number of *genes*. The number of chromosomes in each cell is constant for a particular species. For example, human body cells have the *diploid* number of 46 chromosomes (or 23 *homologous* pairs). *Gametes*, however, contain only half this number (the *haploid* number), which is 23 in humans.

■ *TIP* Try not to confuse the terms chromatid, chromatin and chromosome.

cilia: short, hair-like structures found on the surface membranes of certain cells.

■ Cilia are composed of microtubules and usually occur in large groups on the surface of certain *Protoctista* and on some types of mammalian epithelial cells. They are used for cell movement and to transport substances along a tube.

■ *e.g.* The protoctist **Paramecium** uses cilia for locomotion. Cells lining a *bronchus* in humans use cilia to move mucus containing trapped particles out of the lungs.

■ *TIP* Cilia are frequently confused with *flagella*. Although they are similar in structure, flagella tend to be much longer and usually only one is present on a cell.

circadian rhythm: a 24-hour cyclical change in the behaviour or physiology of organisms.

■ *e.g.* Leaf movements in plants and sleep/activity cycles in many animals.

classification: the system by which living organisms are organised into groups.

■ The most useful classification system is based on the evolutionary relationships between organisms and has a hierarchical structure. All organisms belong to one of five *kingdoms*, each of which can be subdivided into a number of phyla. Each *phylum* can also be divided into classes, and so on. The smallest group is an individual *species*. The table below shows how humans are classified.

Kingdom	Animalia
Phylum	Chordata
Class	Mammalia
Order	Primates
Family	Hominidae
Genus	*Homo*
Species	*sapiens*

■ *TIP* The correct order of the various classification groups is: kingdom, phylum, class, order, family, genus, species. Try to devise a suitable memory-aid to help you learn this sequence, such as Keep Plucking Chickens Or Fear Getting Sacked.

clone: any of the genetically identical individuals that have arisen from a single parent by *asexual reproduction*.

■ Cloning is important in plant breeding and *gene technology*, where it is useful to produce large numbers of identical individuals for commercial purposes.

Cnidaria: the animal phylum containing hydra, jellyfish, sea anemones and corals.

■ Cnidaria share the following features:
  • two layers of cells in the body wall (diploblastic), separated by a jelly-like mesogloea

- radially symmetrical
- tentacles with stinging cells called nematocysts
- a sac-shaped body cavity (enteron) with a single opening acting as a mouth and anus

CNS: see *central nervous system.*

codominant alleles: any two *alleles* which are both expressed in the *phenotype* of a *heterozygous* organism.

■ *e.g.* The gene C for coat colour in cattle has two codominant alleles, C^R (red hairs) and C^W (white hairs). A heterozygous individual ($C^R C^W$) will have a mixture of red and white hairs, giving an overall colour known as roan.

codon: a triplet of bases in *messenger RNA (mRNA)* that codes for a particular *amino acid* during the synthesis of proteins.

■ *e.g.* AUG codes for 'start' in *translation* and the amino acid methionine. GUG codes for the amino acid valine. UAG codes for 'stop' in translation.

■ *TIP* Do not confuse this term with *anticodon*, which is a triplet of bases in *transfer RNA (tRNA)*.

coelom: a fluid-filled cavity formed in the mesoderm of *triploblastic* animals.

■ The importance of the coelom is that it allows the development of specialised organs for both locomotion and digestion. The coelom may also act as a hydrostatic skeleton, as in earthworms, or a circulatory system, as in insects.

coenzyme: an organic molecule that associates with an *enzyme* to catalyse a biochemical reaction.

■ *e.g.* *NAD* is a coenzyme for dehydrogenase enzymes in *respiration*.

cohesion–tension theory: a theory used to explain the movement of water through the *xylem* vessels of a plant.

■ The theory is summarised in the diagram below.

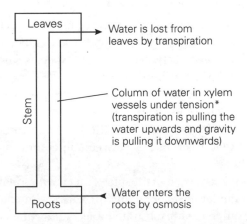

*The water molecules are held together by **cohesion** (forces of attraction between the molecules) and the column is under **tension**. Water molecules are also attracted to the walls of the xylem vessels (adhesion) and this helps to pull the water upwards.

C

collagen: a fibrous protein found in the connective tissue of skin, tendons, *cartilage* and *bone*.

■ Each collagen molecule consists of three polypeptide chains wound into a helix. These helices are bound together to form strong fibres which are very resistant to stretching.

collecting duct: the final section of a kidney *nephron*.

■ The collecting ducts are the main site of water reabsorption in the kidney. The cells lining the collecting ducts are normally relatively impermeable to water. When the body is dehydrated, however, *antidiuretic hormone (ADH)* increases the permeability of these cells and so increases water reabsorption (resulting in the production of a small quantity of concentrated urine).

collenchyma: a type of plant tissue in which the cells have additional *cellulose* thickening in their walls.

■ The main function of collenchyma is to provide mechanical support and it is most commonly found in the stem and leaves.

colon: part of the large intestine, found between the small intestine and the rectum.

■ The main function of the colon is to absorb water and mineral salts from indigestible food material, resulting in the production of faeces.

commensalism: a relationship between two different *species* which results in benefit to one (the commensal) without affecting the other (the host).

■ *e.g.* Remoras are fish that attach themselves to sharks and ride along to pick up scraps of food when the sharks feed. The remoras (commensals) gain food, but the sharks (hosts) appear to be unaffected by the relationship.

community: all the living organisms present in a *habitat*.

■ A climax community represents the different *species* of organisms present at the end of a *succession*, such as oak woodland in many parts of the UK.

competition: the interaction between two or more organisms seeking the same environmental resources that are in short supply, such as food or mates.

■ *Interspecific competition* occurs between different species and *intraspecific competition* takes place between members of the same species. Organisms that compete successfully are more likely to survive and reproduce. Competition is therefore an important force in evolution.

■ *e.g.* Plants often compete for the available light. Predators compete for prey.

■ *TIP* Try to remember that 'inter' means between different things and 'intra' means within something.

competitive inhibition: a reduction in the rate of an enzyme-catalysed reaction by a molecule that is similar in shape to the normal substrate.

■ Inhibitor molecules compete with substrate molecules for the *active site* of an *enzyme*. The extent of the inhibition depends upon the relative concentrations of inhibitor and substrate.

■ *TIP* Try not to confuse this term with *non-competitive inhibition*, which is unaffected by changes in substrate concentration.

condensation reaction: a chemical reaction in which two molecules are joined together and a molecule of water is removed.

■ Condensation reactions are very important in the formation of biologically important polymers or macromolecules.

■ *e.g.* *amino acids* combine to form *polypeptides*.
Monosaccharides combine to form *polysaccharides*.
Glycerol and *fatty acids* combine to form *triglycerides*.

■ *TIP* The reverse of condensation is *hydrolysis*, the breakdown of polymers by the addition of water molecules.

cone cell: a light-sensitive cell found in the *retina* of the eye.

■ Cone cells are responsible for colour vision and function most effectively in bright light. They are densely packed into one area of the retina, known as the *fovea*. This close packing, together with the fact that each cone is connected to a separate sensory *neurone*, means that the eye is able to see very fine detail.

■ *TIP* A common examination question asks you to compare cone cells with *rod cells* in the retina.

Coniferophyta: the plant *phylum* which contains pines, firs and other conifers.

■ Coniferophyta share the following features:
  ● they are usually trees with needle-like leaves
  ● they produce cones rather than flowers
  ● the *seeds* are not enclosed in fruits

connective tissue: an animal tissue consisting of a number of cells embedded in a matrix of protein fibres and salts.

■ The most common protein fibres in connective tissue are *collagen* and elastin (a fibrous protein that is elastic in nature). Connective tissue has many functions, including support, insulation and protection against bacterial invasion.

■ *e.g.* Adipose tissue acts as a thermal insulator and energy store. *Bone* provides support for an organism and protects vital organs. *Cartilage* functions as a shock absorber at the ends of bones in *synovial joints*.

conservation: the process of maintaining an *ecosystem* in order to retain maximum *species* diversity.

■ Conservation involves careful use of the earth's natural resources in order to avoid the gradual destruction of the environment by human activities. It also involves preservation of natural habitats and an awareness of the dangers of *pollution*.

consumer: any of the *heterotrophic* organisms in a *food chain* or *food web*.

■ Herbivores feed on *producers* (green plants) and are known as primary consumers. Carnivores may be secondary consumers (if they feed on herbivores), or tertiary consumers (if they feed on other carnivores).

continuous variation: variation of a characteristic within a population, such that a complete range of forms can be seen.

■ Height in humans is a good example of continuous variation. Within limits, any height is possible, as shown in the graph below.

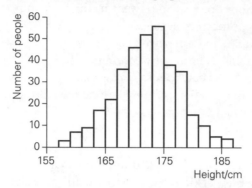

Continuous variation usually occurs when characters are controlled by a number of different genes (polygenes) and where the environment plays a significant role.

■ *TIP* Try not to confuse this term with *discontinuous variation*, where individuals can be assigned to discrete categories, such as ABO *blood groups*.

contractile vacuole: an *organelle* found in some single-celled organisms which removes excess water from the cell.

■ Contractile vacuoles are common in freshwater *Protoctista*, such as **Amoeba**. Water moves into the cell by *osmosis* and must be removed or the organism will burst. The contractile vacuole gradually fills with the excess water and then discharges its contents out through the cell surface membrane. This is an active process and requires *ATP*.

corolla: a collective name for the petals of a *flower*.

■ The corolla is found inside the calyx (a ring of *sepals*) and surrounds the reproductive organs of the plant. It is often brightly coloured in plants pollinated by insects, but much reduced or absent in wind-pollinated plants.

coronary heart disease: the blockage of one or more of the coronary *arteries* which supply blood to the muscle of the *heart*.

■ Blockages occur in the coronary arteries due to the build-up of fatty material (atheroma) in the lining of the artery wall, known as atherosclerosis. A blood clot can also block one of the arteries and this is known as thrombosis. Blockage of a coronary artery will prevent oxygen reaching an area of heart muscle. This muscle then dies and causes a myocardial infarction (heart attack).

corpus luteum: a tissue that forms in the *ovary* of a mammal following the release of an *ovum*.

■ The corpus luteum develops from an ovarian follicle after the ovum is released. Its main function is to produce the hormone *progesterone*, which is responsible for maintaining the lining of the *uterus*. The activity of the corpus luteum depends on whether *fertilisation* occurs. If fertilisation does not occur the corpus luteum breaks down after about two weeks. Progesterone levels fall and the

C

lining of the uterus breaks down and is lost from the body during menstrua-tion. If fertilisation occurs the corpus luteum continues to produce proges-terone for about three months. Production of progesterone is then taken over by the *placenta*.

correlation: a relationship between two variables such that a change in one of them is reflected by a change in the other.

■ There are three possible relationships between the variables *X* and *Y*:
- positive correlation — an increase in *X* is accompanied by an increase in *Y*
- negative correlation — an increase in *X* is accompanied by a decrease in *Y*
- no correlation — there is no obvious relationship between *X* and *Y*

These possible relationships are shown as scattergrams in the diagram below.

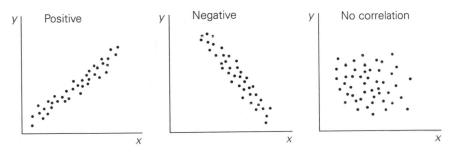

Care must be taken when interpreting scattergrams. Just because two things are correlated, it does not mean that one causes the other. Furthermore, some variables may have a very complex interrelationship and simple scattergrams are only really useful for detecting linear (straight-line) relationships.

cotyledon: the seed leaf of a plant *embryo*.

■ Cotyledons act as a store of food for the developing plant. In seeds without an *endosperm*, such as the broad bean, the cotyledons store food for use during *germination*. In other seeds, such as the runner bean, they emerge above the soil surface and become the first photosynthetic leaves. The number of cotyledons can be used in the *classification* of flowering plants. *Monocotyledons* have a single cotyledon in each seed and *dicotyledons* have two cotyledons in each seed.

counter-current system: a mechanism by which fluids in adjacent systems flow in opposite directions in order to maintain a diffusion gradient between them.

■ *e.g.* Blood flows in the opposite direction to water passing over the gills of fish. This maintains a diffusion gradient for the exchange of respiratory gases (oxygen and carbon dioxide).

crossing-over: the exchange of genetic material between *homologous chromosomes* during *meiosis*.

■ Crossing-over occurs when *chiasmata* form between *chromatids*. This process results in new combinations of *alleles* and therefore produces genetic *variation*.

cultural evolution: the changes which have taken place in human societies over the course of time.

■ Over the last 100 000 years, human communities have evolved from small groups of hunter-gatherers to modern complex societies.

■ *e.g.* There have been several significant steps in the cultural evolution of humans, including tool-making, the control of fire and the development of agriculture.

cyclic AMP (cAMP): a molecule that acts as a second messenger in many physiological reactions induced by *hormones* (first messenger).

■ The diagram below summarises the role of cyclic AMP in hormonal control.

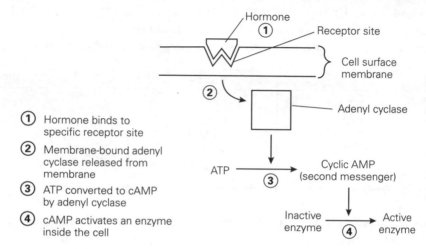

① Hormone binds to specific receptor site

② Membrane-bound adenyl cyclase released from membrane

③ ATP converted to cAMP by adenyl cyclase

④ cAMP activates an enzyme inside the cell

cytochrome: an iron-containing protein which forms part of an *electron transport chain*.

■ Cytochromes are *coenzymes* found on the membranes inside *mitochondria* and *chloroplasts*. They are important in the electron transport chains involved in *respiration* and *photosynthesis*.

cytokinin: a plant growth substance which stimulates cell division.

■ Cytokinins also promote the growth of lateral buds and leaves in stems, and delay senescence (ageing) in leaves.

■ *TIP* Make sure that you know how cytokinins interact with other plant growth substances, such as *abscisic acid*, *auxins*, *ethene* and *gibberellins*.

cytoplasm: the contents of a cell surrounding the *nucleus* and enclosed by the cell surface membrane.

■ Cytoplasm consists of *organelles* suspended in a jelly-like matrix (cytosol) which contains enzymes and numerous small molecules, such as *amino acids* and *nucleotides*. The cytoplasm is metabolically active, being the site of *glycolysis* and *fatty acid* synthesis.

cytosine: a *pyrimidine* nitrogenous base that is found in *DNA* and *RNA*.

■ In DNA, cytosine always forms a bond with *guanine*, a *purine* nitrogenous base.

deamination: the removal of an amino (NH_2) group from an *amino acid*.

■ Excess amino acids are too toxic for storage in the body. Therefore, they are deaminated in the *liver*, forming *carbohydrates* and ammonia. In mammals, the carbohydrate portion is usually respired and ammonia passes into the *ornithine cycle* to be converted to *urea*.

■ *TIP* Because urea is excreted by the kidneys, many students think that deamination occurs in the kidneys and not in the liver. Try to avoid this common mistake.

decomposer: an organism that breaks down dead material or waste products.

■ Decomposers and *detritivores* play an important role in the recycling of nutrients incorporated in the waste or dead remains of plants and animals. Detritivores tend to break up the dead material, providing a large surface area for the action of decomposers.

■ *e.g.* Many species of *bacteria* and *fungi* are decomposers.

deflected succession: a type of *succession* in which a climax *community* fails to become established due to human influences.

■ *e.g.* Using grassland for sheep farming will prevent it developing into woodland (the natural climax community).

denitrification: the process by which *bacteria* covert nitrates in the soil to nitrogen gas in the atmosphere.

■ Denitrifying bacteria, such as ***Pseudomonas denitrificans***, use nitrates as a source of energy and play an important role in the *nitrogen cycle*. The bacteria are particularly active in waterlogged or poorly aerated soils, where the oxygen content is low. Denitrification reduces the quantity of nitrate ions that is available to plants.

■ *TIP* The reactions involved in denitrification are essentially the opposite of those that comprise *nitrification*, as summarised by the *nitrogen cycle*.

density-dependent factor: any factor limiting the size of a *population* whose effect is proportional to the density of the population.

■ The density of a population may be taken as the number of individuals per unit area.

■ *e.g.* Food supply and infectious disease are both density-dependent factors.

■ *TIP* Make sure you are clear about the difference between density-dependent factors (which are usually *biotic*) and *density-independent factors* (which are usually *abiotic*).

density-independent factor: any factor limiting the size of a *population* whose effect is independent of the density of the population.

■ The density of a population may be taken as the number of individuals per unit area.

■ *e.g.* Temperature and water-availability are both density-independent factors.

■ *TIP* Make sure you are clear about the difference between density-independent factors (which are usually *abiotic*) and *density-dependent factors* (which are usually *biotic*).

deoxyribonucleic acid: see *DNA*.

detritivore: an animal that feeds on dead material or waste products.

■ Detritivores and *decomposers* play an important role in the recycling of nutrients incorporated in the waste or dead remains of plants and animals (detritus). Detritivores tend to break up the dead material first, providing a large surface area for the action of decomposers.

■ *e.g.* Earthworms and woodlice are detritivores.

diabetes: a disease in which the concentration of *glucose* in the blood cannot be properly regulated.

■ There are two types of diabetes.

- Insulin dependent diabetes: the pancreas fails to produce enough *insulin* to control the concentration of glucose in the blood. Excess glucose appears in the *urine*, leading to weight loss and dehydration. The disease can be controlled by regulating the diet and injecting insulin.

- Non-insulin dependent diabetes: the pancreas produces enough insulin, but the cells of the body fail to take up glucose in response to the hormone. The disease can only be controlled by carefully regulating the diet.

dialysis: a method for separating molecules of different sizes by diffusion through a partially permeable membrane.

■ The process of dialysis takes place naturally during *ultrafiltration* in mammalian kidneys.

■ *e.g.* If a mixture of starch and glucose is placed into a container as shown below, the smaller glucose molecules will pass through the membrane into the water while the larger starch molecules remain behind.

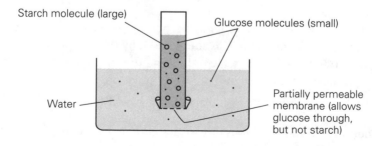

Starch molecule (large)
Glucose molecules (small)
Water
Partially permeable membrane (allows glucose through, but not starch)

d

dicotyledon: a flowering plant that produces seeds with two *cotyledons* (seed leaves).

■ Dicotyledons share the following features:
- a net-like pattern of veins in the leaves
- vascular bundles arranged in a ring in the stem
- flowers in four or five parts, or multiples of these numbers

■ *e.g.* Buttercups and oak trees are dicotyledons.

■ *TIP* Make sure that you know the differences between dicotyledons and *mono-cotyledons*, which produce seeds with only one cotyledon.

differentiation: the process by which cells become specialised for a particular function.

■ *e.g.* Simple cells in the root tip of plants undergo differentiation to form specialised tissues, such as *xylem* and *phloem*.

diffusion: the net movement of *molecules* or *ions* from a region of high concentration to a region of low concentration.

■ Diffusion is a passive process (it does not require energy) and the random movement of molecules or ions will continue until they are uniformly distributed. The rate of diffusion is increased by any of the following factors:
- a large concentration gradient
- a small diffusion distance
- higher temperatures
- a membrane that is fully permeable to the molecules or ions
- small molecules or ions

Likewise, the rate of diffusion will be decreased if any of these factors do not apply.

■ *e.g.* Respiratory gases (oxygen and carbon dioxide) diffuse in and out of cells.

digestion: the breakdown of food material into simple molecules that can be absorbed by the body.

■ Digestion in a mammal involves both physical and chemical processes. Physical digestion refers to the action of teeth and gut muscles in helping to break up food and mix it with digestive juices. Chemical digestion occurs when *bile* breaks down fat into tiny droplets and hydrolytic enzymes complete the process of digestion by breaking down the individual food molecules.

■ *e.g.* *Lipase*, a digestive enzyme produced by the pancreas, hydrolyses *lipids* into *fatty acids* and glycerol.

■ *TIP* There are a number of important digestive enzymes and you should know where each one is produced, its substrates and products, and the conditions under which it works best.

dihybrid inheritance: the inheritance of two different characteristics.

■ The two characteristics are controlled by *genes* at different loci (positions on a *chromosome*). If the genes are on the same chromosome, they will show autosomal *linkage*. If, however, the genes are on different chromosomes, they will show independent assortment.

d

■ *e.g.* A dihybrid cross between two pure-breeding pea plants, one with smooth yellow seeds and one with wrinkled green seeds, would be expected to produce an F_1 *generation* with only smooth yellow seeds. This is because smooth texture and yellow coloration are dominant, and wrinkled texture and green coloration are recessive. If these F_1 plants are interbred (and assuming the genes for seed texture and colour are on different chromosomes), the F_2 generation will have a mixture of seeds: smooth yellow, smooth green, wrinkled yellow and wrinkled green; in a ratio of 9:3:3:1. This dihybrid cross is summarised in the diagram below.

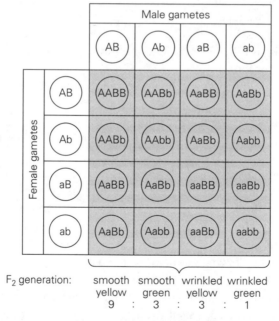

Let the alleles be represented by the following letters:
A = smooth; B = yellow; a = wrinkled; b = green

Parents: Pure-breeding plants with × Pure-breeding plants with
smooth yellow seeds wrinkled green seeds
AABB aabb

Gametes: AB ab

F_1 generation: AaBb
(smooth yellow seeds)

Interbreeding the F_1 generation:

Male gametes
AB Ab aB ab

Female gametes
AB | AABB AABb AaBB AaBb
Ab | AABb AAbb AaBb Aabb
aB | AaBB AaBb aaBB aaBb
ab | AaBb Aabb aaBb aabb

F_2 generation: smooth smooth wrinkled wrinkled
yellow green yellow green
9 : 3 : 3 : 1

■ *TIP* It is a good idea to practise writing out genetic crosses, as in the examples above. Try to remember the following points.

• Set your work out clearly, labelling the parents, gametes and new generations.
• Do not confuse the terms *genotype* and *phenotype*, or *gene* and *allele*.
• Do not represent alleles with letters that look similar in capitals and lower case, such as S and s, or K and k.

- Remember that gametes are *haploid* — they only have one of each pair of alleles.
- Be aware that a ratio of 9:3:3:1 (or any other genetic ratio) is only what we would expect to find. The actual numbers of the different organisms may often be slightly different. If you do not find the ratio you expect, it may be a case of *autosomal* or *sex linkage*.

diploblastic: an animal with a body wall composed of only two layers of cells.

■ Members of the *phylum Cnidaria*, such as jellyfish, are diploblastic. They possess two layers of cells, the endoderm (inner layer) and the ectoderm (outer layer), separated by a jelly-like mesogloea. Most other animals are *triploblastic*.

diploid: a cell in which the *nucleus* contains two complete sets of *chromosomes*.

■ Most body cells in humans are diploid (sometimes referred to as 2n), possessing 46 chromosomes in the nucleus. In other words, the diploid number for humans is 46. *Gametes* are an exception to this rule, as they have only half the number of chromosomes that are present in a body cell. For example, a human sperm cell contains 23 chromosomes, which is known as the *haploid* (n) number.

■ *TIP* Remember, the diploid number of chromosomes is variable, and differs from species to species. For example, the diploid number in pea plants is 14.

directional selection: a type of *natural selection* which occurs when environmental change favours a new form (*phenotype*) of an organism.

■ The process of directional selection is summarised in the diagram below.

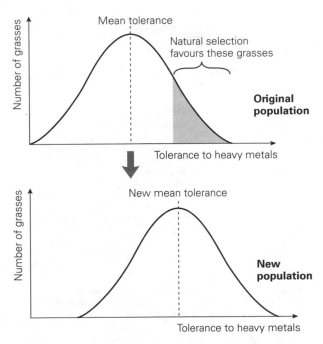

■ *e.g.* In soils contaminated with heavy metals, such as lead, tin, or zinc, natural selection will favour grasses that can tolerate these metals.

■ *TIP* Make sure that you understand the differences between directional selection and the other two forms of natural selection: *disruptive selection* and *stabilising selection*.

disaccharide: a type of *carbohydrate* formed from two *monosaccharides*.

■ Disaccharides are formed by a *condensation reaction* between two hexose (six carbon) monosaccharides, resulting in the formation of a glycosidic bond.

■ *e.g.* Sucrose (formed from glucose and fructose); lactose (formed from glucose and galactose); and maltose (formed from two molecules of glucose).

■ *TIP* The condensation reaction which produces disaccharides can be reversed. For example, the *hydrolysis* of sucrose will break the molecule down into its monosaccharide sub-units, glucose and fructose.

discontinuous variation: a type of *variation* in which there are clearly defined differences within a population.

■ In general, discontinuous characteristics are determined by one or two *genes* and the environment has little or no effect on the final pattern of variation.

■ *e.g.* ABO *blood groups* in humans. These groups are controlled by a single gene, and an individual will either be group A, B, AB, or O, with no intermediate forms.

■ *TIP* Make sure that you understand the differences between discontinuous variation and *continuous variation*, where a complete range of forms can be seen.

disruptive selection: a type of *natural selection* in which the environment favours forms at the extremes of the range of phenotypic variation.

■ The effect of disruptive selection is to eliminate *phenotypes* in the middle of the range, as shown in the diagram below.

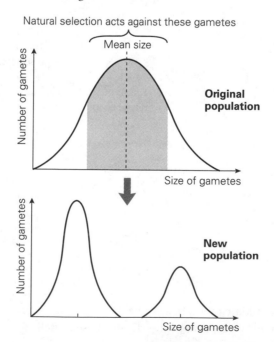

■ *e.g.* Disruptive selection may have occurred during the evolution of specialised *gametes* (eggs and sperm). In this case, small gametes (sperm) could be produced in large numbers, and large gametes (eggs) would have sufficient energy to survive and nourish the initial growth of the *zygote*. Both would have advantages over medium-sized gametes, which would be selected against.

■ *TIP* Make sure that you understand the differences between disruptive selection and the other two forms of natural selection: *directional selection* and *stabilising selection*.

divergent evolution: see *adaptive radiation*.

diversity: a measure of the number of different *species* present in a particular *habitat*.

■ *e.g.* A desert habitat will only have a few species present and therefore has a low diversity. Tropical rainforests, on the other hand, contain a very large number of species, and so have a high diversity.

DNA (deoxyribonucleic acid): a molecule that forms the genetic material of all living organisms.

■ DNA is a *nucleic acid*, consisting of two polynucleotide chains coiled into a double helix. A simplified diagram of part of a DNA molecule is shown below.

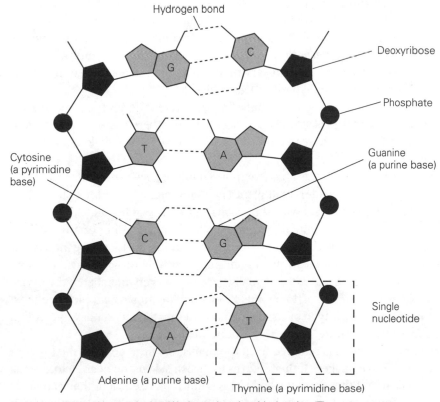

Note: adenine (A) always bonds with thymine (T)
guanine (G) always bonds with cytosine (C)

d

DNA is a major constituent of the *chromosomes* in the nucleus of a *eukaryotic* cell. It is also found in a non-chromosomal form in *prokaryotic* cells, *chloroplasts*, *mitochondria* and *viruses*. DNA plays a central role in the determination of hereditary characteristics by controlling *protein synthesis* in cells. A section of DNA that codes for the production of a particular protein or polypeptide is known as a *gene*.

■ *TIP* Remember that DNA is not a protein. A common examination question asks you to compare the structure of DNA with that of *RNA (ribonucleic acid)*.

DNA probe: a single strand of *DNA* that is used to identify a particular *gene*.

■ A DNA probe is usually radioactively labelled and has a complementary base sequence to the gene to be identified. DNA from the organism in question is fragmented into short sequences, treated to separate the two chains and then incubated with the probe. If the gene is present on one of the DNA fragments, the probe will bind to it and the gene can be identified by its radioactivity. DNA probes are commonly used in the diagnosis of genetic disorders.

DNA replication: the process by which a *DNA* molecule produces two exact copies of itself.

■ DNA replication is controlled by the enzyme DNA polymerase. *Hydrogen bonds* break between the two polynucleotide chains and the parent DNA molecule unwinds. Each DNA strand then acts as a template for the synthesis of a new strand. Free *nucleotides* line up opposite their complementary bases and are joined together by *condensation reactions* to form two new DNA molecules. This process is known as semi-conservative replication, as each new molecule contains half the original parent DNA molecule.

■ *TIP* It is a common mistake to think that DNA replication is the same as *transcription*, which takes place during *protein synthesis*. There are several important differences between these two processes and it is vital that you do not get them confused.

dominant: any *allele* that is always expressed in the *phenotype* of an organism.

■ *e.g.* In peas, the allele for a tall plant (T) is dominant to that for a short plant (t). Therefore, pea plants with the *genotype* TT or Tt will be tall.

■ *TIP* Alleles may be dominant, *recessive* or *codominant*. It is important that you know and understand all of these terms. In addition, just because an allele is dominant, it does not mean that organisms possessing this allele will be more healthy or that the allele will eventually be found in every member of the *species*. For example, a number of dominant alleles found in humans have lethal effects!

double circulation: a type of circulatory system in which blood passes through the heart twice during its journey around the body.

■ In mammals, blood is pumped from the right side of the heart to the lungs in order to receive oxygen. It then returns to the left side of the heart before being pumped around the rest of the body. This is much more efficient than the single circulation system seen in fish, because the tissues are always supplied with oxygen-rich blood at high pressure.

d

Down's syndrome: a genetic disorder in humans caused by the presence of an additional *chromosome*.

■ Most cases of Down's syndrome are due to *non-disjunction* of chromosome 21 during *meiosis*. These *homologous chromosomes* fail to separate, resulting in the production of a *gamete* with an extra copy of chromosome 21. If this gamete is involved in *fertilisation*, the resulting *zygote* will have three copies of chromosome 21 (instead of the normal two copies) and will develop into an individual with Down's syndrome. Some cases of Down's syndrome are due to a translocation *mutation*. In these cases, chromosome 21 becomes attached to chromosome 15. This may result in a gamete containing the mutant chromosome, as well as the usual single copy of chromosome 21. In other words, it will have two copies of chromosome 21. Down's syndrome will then develop, as described in the non-disjunction example given above. The symptoms of Down's syndrome include: mental retardation, heart defects and characteristic facial features.

duodenum: the first section of the small intestine.

■ The duodenum receives partly digested food from the *stomach*. It secretes alkaline mucus to neutralise the stomach acid and provide the correct *pH* for the intestinal *enzymes* to function. *Bile* and pancreatic enzymes are secreted into the duodenum to continue the process of *digestion*. Some of the products of digestion are absorbed in the duodenum, although most *absorption* takes place in the *ileum*.

ECG: see *electrocardiogram*.

Echinodermata: the animal *phylum* containing sea urchins and starfish.

■ Echinoderms share the following features:
- *radial symmetry*
- an *exoskeleton* consisting of calcium carbonate plates under the surface of the skin
- a water vascular system and tube feet for gas exchange, *excretion* and locomotion

ecology: the study of the interactions of organisms with their *biotic* (living) and *abiotic* (non-living) environments.

■ In general, ecologists study organisms in terms of the *ecosystem* in which they are found.

■ *TIP* Ecology questions in examinations often attract very general answers. Make sure that your answers are detailed, including several examples that you have learned on your biology course.

ecosystem: a stable but dynamic system, characterised by the interaction of its *biotic* (living) and *abiotic* (non-living) components.

■ Ecosystems consist of one or more *communities* of organisms living in a particular *habitat*. Each community contains *populations* of *producers*, *consumers* and *decomposers*, and each population occupies a particular ecological *niche* within the habitat.

■ *e.g.* A pond may be considered as a separate ecosystem, containing plants, insects, fish and a variety of decomposers. These living organisms will interact with each other and react to the non-living components of the ecosystem, such as temperature, light intensity and the availability of mineral salts.

ectotherm: an animal that uses the environment to regulate its body temperature.

■ All animals other than birds and mammals are ectothermic and are unable to regulate their body temperature by physiological methods. All ectotherms are therefore subject to temperature fluctuations as the environmental temperature changes. However, many of these animals can regulate their body temperature within certain limits by using behavioural mechanisms. For example, crocodiles gain heat by basking in the sun on land, and lose heat by taking to the water.

e

■ **TIP** Try not to confuse this term with *endotherm*, which refers to an animal that can regulate its body temperature using physiological mechanisms.

effector: a cell or organ that responds to a *stimulus*.

■ In animals, *muscles* and *glands* are both effectors that respond to stimulation by a nerve *impulse* or a *hormone*. A typical sequence of events involving a response to a stimulus is shown below.

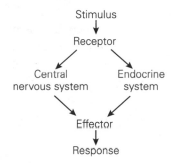

electrocardiogram (ECG): a graph showing the electrical activity in the heart during the *cardiac cycle* (heartbeat).

■ A typical ECG for a healthy adult is known as a PQRST wave, as shown in the diagram below.

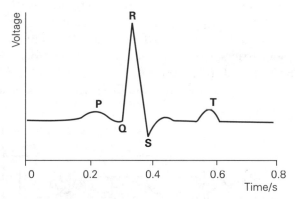

The small peak at P represents atrial systole (the contraction of the *atria*), the QRS complex represents ventricular systole (the contraction of the *ventricles*) and the peak at T represents atrial and ventricular diastole (relaxation of the atria and ventricles). Changes from this normal pattern may indicate heart disease.

electron: a negatively charged particle normally found orbiting the nucleus of an *atom*.

■ Electrons are important in making bonds with other atoms to form *molecules*. If the number of electrons is equal to the number of protons (positively charged particles in the nucleus), then the atom or molecule will have no charge. Otherwise it will be a negatively charged *ion* (more electrons than protons) or a positively charged ion (more protons than electrons). Free electrons are important in the *electron transport chain*.

■ **TIP** *Oxidation* and *reduction* reactions are important in biochemistry. Oxidation is the gain of oxygen, the loss of hydrogen, or, more accurately, the loss of electrons. Reduction is the loss of oxygen, the gain of hydrogen, or, more accurately, the gain of electrons. To help you remember this, think OILRIG (Oxidation Is Loss Reduction Is Gain).

electron microscope: a microscope that uses a beam of electrons to form highly magnified images of very small objects.

■ An electron microscope is very similar to a light microscope, except that it uses an electron beam instead of light and powerful electro-magnets instead of lenses. As a result, the electron microscope provides resolution up to 400 times better than a light microscope. This has enabled biologists to study the structure of cell *organelles*, *viruses* and large molecules, such as *DNA*. It should be noted, however, that electron microscopes are very expensive to operate and can only be used to study non-living material.

■ **TIP** A common examination question asks you to compare the advantages and disadvantages of electron and light microscopes.

electron transport chain: a system of carrier molecules which transfer *electrons* from one to the other, releasing energy for the production of *ATP*.

■ Electron transport chains can be found on the cristae of *mitochondria*, where they carry out the last stage of *aerobic respiration*, and on the *thylakoid* membranes of *chloroplasts*, where they generate ATP during the *light-dependent stage* of *photosynthesis*. In both systems, the energy released by the transfer of electrons between carriers is used to pump protons (hydrogen *ions*) across membranes. The return of the protons, through specialised protein pores in the membranes, is coupled to the synthesis of ATP. This process is summarised in the diagram below.

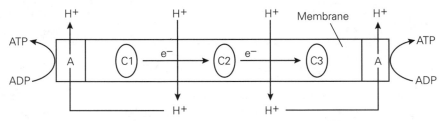

Key: H+ = proton; C1/C2/C3 = electron carriers; e− = electron; A = ATPase

electrophoresis: a method used to separate a mixture of charged molecules.

■ In biology, electrophoresis is mainly used to analyse mixtures of *proteins* or *nucleic acids*. These molecules will migrate in an electric field, with negatively charged molecules moving towards the positive electrode and positively charged molecules moving towards the negative electrode. The rate of movement depends upon the size, shape and overall charge of the molecules.

element: any substance that cannot be broken down further by chemical means.

■ Carbon (C), hydrogen (H), oxygen (O) and nitrogen (N) are all elements. Combinations of atoms of the same or different elements are called *molecules*, such as carbon dioxide (CO_2), oxygen gas (O_2) or *glucose* ($C_6H_{12}O_6$).

embryo: an animal or plant that develops from a *zygote* prior to birth, hatching or *germination*.

■ In flowering plants, the embryo is contained within the *seed* and consists of the embryonic shoot (*plumule*) and root (*radicle*), and either one or two seed leaves (*cotyledons*). In animals, the embryo develops inside an egg or, in mammals, in the *uterus*. A human embryo is usually called a *fetus* after the first eight weeks of pregnancy.

emulsion test: a *biochemical test* that can be used to show the presence of a *lipid*.

■ The test involves shaking the sample with ethanol and then pouring the resulting solution into a test-tube containing water. A white emulsion indicates that lipid is present.

endocrine gland: a gland that secretes a *hormone* directly into the blood.

■ *e.g.* The *pancreas* secretes *insulin* and *glucagon* into the blood.

■ *TIP* Make sure that you understand the difference between endocrine (ductless) glands and *exocrine glands*, which secrete substances through a duct to their place of action.

endocytosis: the uptake of large particles or fluids through the surface membrane of a cell.

■ During endocytosis, the cell surface membrane surrounds the substance in question and forms a *vesicle* or *vacuole*, which moves into the *cytoplasm*. The terms *phagocytosis* and *pinocytosis* are sometimes used to refer to the uptake of solid particles and fluids respectively.

■ *TIP* Try not to confuse this term with *exocytosis*, which refers to the movement of substances out of cells, across the cell surface membrane. In general, 'endo' means 'into' or 'within', and 'exo' or 'ecto' means 'out of' or 'outside'.

endodermis: a layer of cells surrounding *vascular* tissue in the *root* of a plant.

■ Endodermal cells possess a band of impermeable *suberin* known as the *Casparian strip*. This prevents water and solutes from entering the *xylem* via the *apoplastic pathway* (through the cell walls and intercellular spaces). As a result, these substances have to pass along the *symplastic pathway* (through the *cytoplasm* of the endodermal cells). This allows the plant to control the movement of water and solutes into the xylem.

endometrium: the inner lining of the *uterus* of mammals.

■ The endometrium provides nutrients for the young *embryo* before the *placenta* is established. In the absence of pregnancy, the periodic development and removal of the endometrium represents the *menstrual cycle*.

endopeptidase: an *enzyme* that breaks *peptide bonds* at specific sites within a protein molecule.

■ Endopeptidases tend to produce small *polypeptides* and very few single *amino acids*. These small polypeptides are then digested by *exopeptidases*, which are enzymes that remove amino acids and dipeptides (two amino acids joined together) from the ends of a polypeptide chain.

e

■ *e.g. Trypsin* is an endopeptidase that breaks a polypeptide chain next to basic amino acids, such as lysine.

endoplasmic reticulum: a system of membranes found in the *cytoplasm* of a cell.

■ The endoplasmic reticulum (ER) provides a very large surface area for chemical reactions within the cell. It also acts as a pathway for the transport of materials and provides structural support for the cell. There are two types of ER in cells.

- Rough ER has an outer surface lined with *ribosomes* and is involved in the synthesis of proteins.
- Smooth ER has no ribosomes and is concerned with the synthesis of *lipids* and *steroids*.

endoskeleton: the supportive internal framework of an organism.

■ Vertebrates have a well-developed endoskeleton, consisting of *bone* and *cartilage*. The endoskeleton supports the body, protects the organs and works with muscles to produce movement.

■ *TIP* Try not to confuse this term with *exoskeleton*, which performs similar functions in insects, but is found on the outside of the body.

endosperm: a nutritious tissue found in the *seeds* of flowering plants.

■ The endosperm provides nutrients for the developing *embryo* within the seed. In *dicotyledons*, the endosperm is absorbed during the development of the seed. In *monocotyledons*, however, there is still a significant quantity of endosperm present in the mature seed. In these cases, the endosperm acts as a food store during *germination*.

endotherm: an animal that can regulate its body temperature using physiological mechanisms.

■ Birds and mammals are endothermic. This means that they can generate heat within the body and maintain a constant temperature despite fluctuations in the environmental temperature. The diagram below summarises the regulation of body temperature in humans.

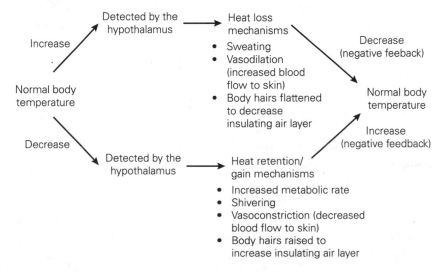

Temperature regulation in endotherms is a good example of *negative feedback*. A change in body temperature stimulates various mechanisms that return the temperature to normal.

■ *TIP* Try not to confuse this term with *ectotherm*, which refers to an animal that cannot regulate its body temperature using physiological mechanisms.

energy: see *pyramid of numbers/biomass/energy*.

enterokinase: an *enzyme* secreted in the small intestine that converts trypsinogen to *trypsin*.

■ Trypsin is an *endopeptidase*, which is secreted in an inactive form known as trypsinogen. By activating this enzyme, enterokinase plays a key role in the digestion of proteins.

environment: the conditions found in the region in which an organism lives.

■ The environment can be divided into *abiotic* (non-living) factors, such as temperature, and *biotic* (living) factors, such as food availability.

enzyme: a protein molecule which acts as a catalyst in living organisms.

■ Enzymes are globular proteins that speed up biochemical reactions by lowering the *activation energy* required for the reactions to take place. The individual reactions that constitute *metabolism* are catalysed by specific enzymes, which act according to the *lock and key hypothesis* or the *induced fit hypothesis*. Essentially, enzymes combine with a specific *substrate* and convert it to a product according to the equation below.

$$E + S \rightleftharpoons ES \rightleftharpoons EP \rightleftharpoons E + P$$

Enzyme + substrate	Enzyme–substrate complex	Enzyme–product complex	Enzyme + product

Several factors influence the rate of enzyme-controlled reactions. These include: the concentration of enzyme and substrate; temperature; *pH*; and the presence of *coenzymes*, activators or inhibitors.

■ *e.g.* Lipase is an enzyme secreted by the *pancreas* into the small intestine, where it hydrolyses *lipids* (the substrate) into *fatty acids* and glycerol (the products).

■ *TIP* Examination questions will often ask you to explain the influence of a certain factor on the rate of an enzyme-controlled reaction. Make sure that you do not simply describe the effect if an explanation is needed. Remember that excessive heat denatures enzymes by disrupting their tertiary structure; it does not kill them! Low temperatures do not denature enzymes, they simply work more slowly because there is less energy in the system.

epidermis: the outer layer of cells in a plant or animal.

■ Epidermal cells often have a protective function. For example, the epidermis in plants protects the delicate tissues beneath and also prevents excessive water loss. Similarly, in mammals the epidermis consists of several layers of skin cells, the outer ones of which contain keratin (a tough structural protein) and are dead.

epistasis: a type of *gene* interaction, in which one gene controls the expression of another gene.

■ *e.g.* The fur colour of guinea pigs is controlled by two genes. If an animal has a dominant *allele* A it will have coloured fur; otherwise it will be albino (white). If it also has dominant allele B it will have black fur; if not, its fur will be brown. In other words: all animals without A will be white; those with A and not B will be brown; and those with A and B will be black. The possible *genotypes* and *phenotypes* of these guinea pigs are summarised in the table below.

Genotypes	Phenotype (fur colour)
AABB, AaBB, AABb, AaBb	Black
AAbb, Aabb	Brown
aaBB, aaBb, aabb	White

epithelium: a layer of closely packed cells which forms a covering around body structures.

■ There are three basic types of epithelium:

- squamous (pavement) epithelium — flattened cells found in the *alveoli* of the lung
- cuboidal epithelium — cube-shaped cells found in the kidney tubules
- columnar epithelium — elongated cells found in the small intestine

Epithelium has a number of functions, including the protection of tissues, *absorption* of materials and secretion of useful substances. All substances that enter or leave body tissues must pass through epithelial cells. As a result, these cells are often specialised for the exchange of materials.

ethene: a gaseous hydrocarbon (C_2H_4) which acts as a *plant growth substance*.

■ Ethene stimulates the ripening of *fruit* and, in some species, breaks the dormancy of buds and *seeds*.

■ *TIP* Make sure that you know how ethene interacts with other plant growth substances, such as *abscisic acid, auxins, cytokinins* and *gibberellins*.

eukaryotic: cells with a distinct *nucleus* containing a number of *chromosomes*.

■ All living organisms except *bacteria* are eukaryotic. In addition to a nucleus, the cells also contain membrane-bound *organelles*, such as *mitochondria* and *chloroplasts*.

■ *TIP* A common exam question asks you to compare eukaryotic cells with *prokaryotic cells*, such as bacteria. Make sure you are clear about the similarities and differences between these two types of cells.

eutrophication: a decrease in biodiversity resulting from the pollution of a river or lake by sewage or *fertilisers*.

■ The flow-chart below summarises the process of eutrophication in a lake.

Fertiliser or sewage enters lake

↓

Increased nitrate and phosphate concentration

↓

Stimulates growth of algae (algal bloom)

↓

Algae eventually die, leading to an increase
in the respiration of aerobic decomposers

↓

Increased biochemical oxygen demand
(BOD) and subsequent reduction in
the oxygen concentration of the lake

↓

Death of many organisms (e.g. fish)
due to lack of oxygen

■ *TIP* Note that it is not the algal bloom that reduces the oxygen concentration of the water, but the increase in aerobic respiration by the decomposing bacteria.

evolution: the process by which new *species* arise as the result of gradual change to the genetic make-up of existing species over long periods of time.

■ Charles Darwin's theory of evolution by *natural selection* can be summarised as follows. All species tend to produce many more offspring than can ever survive. However, the size of populations tends to remain more or less constant, meaning that most of these offspring must die. It also follows that there must be competition for resources such as mates, food and territories. Therefore, there is a 'struggle for existence' among individuals. Individuals within a species differ from one another (*variation*) and many of these differences are inherited (genetic variation). Competition for resources, together with variation between individuals, means that certain members of the population are more likely to survive and reproduce than others. These individuals will inherit the characteristics of their parents and evolutionary change will take place through natural selection. Over a long period of time, this process may lead to the considerable differences now observed between living organisms.

■ *TIP* A common examination question asks about the evolution of resistance to *antibiotics* in *bacteria*. Remember that antibiotics do not cause these bacteria to mutate. The *mutation* (gene for resistance) already exists and the use of antibiotics simply selects those bacteria that carry this mutation. The resistant bacteria then reproduce and over a period of time the whole population will become resistant.

excretion: the removal of the waste products of cellular *metabolism*.

■ Plants produce few excretory products, as they generally only synthesise what they require. Excess carbon dioxide or oxygen, from *respiration* and *photosynthesis* respectively, diffuses out of the plant via *stomata* and *lenticels*. Other excretory products, such as tannic acid, accumulate in the leaves and are removed when the leaves fall. The main excretory products in mammals are carbon dioxide from cellular respiration and *urea* from the *deamination* of excess *amino acids* in the liver. Excretion of carbon dioxide occurs in expired air and the removal of urea takes place in the kidneys.

■ *TIP* Defaecation (the removal of faeces) is not the same as excretion, because the waste materials are not the products of cellular metabolism. Try not to confuse excretion with secretion, which is the transport of useful substances to their place of action.

exocrine gland: a gland that secretes substances through a duct to their place of action.

■ *e.g.* The *pancreas* secretes digestive *enzymes* into the small intestine via the pancreatic duct.

■ *TIP* Make sure that you understand the difference between exocrine glands and *endocrine glands*, which do not have ducts. Note that the pancreas is both an exocrine and an endocrine gland.

exocytosis: the transport of large particles or fluids out of a cell via the cell surface membrane.

■ *Vesicles* are formed from the *Golgi apparatus* in the *cytoplasm* of the cell. These fuse with the surface membrane, releasing their contents outside the cell.

■ *e.g.* Milk is secreted from the cells of the mammary glands by exocytosis.

■ *TIP* Try not to confuse this term with *endocytosis*, which refers to the movement of substances into cells, across the cell surface membrane. In general, 'endo' means 'into' or 'within', and 'exo' or 'ecto' means 'out of' or 'outside'.

exopeptidase: a protein-digesting *enzyme* that breaks *peptide* bonds at the end of a polypeptide chain.

■ Exopeptidases remove single *amino acids* or dipeptides (two amino acids joined together) from the short polypeptide chains produced by *endopeptidases*.

exoskeleton: the supportive external framework of an organism.

■ Animals in the *phylum Arthropoda* have a well-developed exoskeleton. This is composed mainly of *chitin*, a nitrogen-containing *polysaccharide*, further strengthened by proteins or calcium salts. The exoskeleton supports the body, protects the organs and works with the muscles to produce movement.

■ *TIP* Try not to confuse this term with *endoskeleton*, which performs similar functions in vertebrates, but is found on the inside of the body.

extracellular digestion: the breakdown of food molecules that takes place outside cells.

■ Most of the *digestion* that takes place in the gut of a mammal is extracellular. Food is broken down in the *lumen* of the gut and the products of digestion are then absorbed by the intestinal cells. Digestion that takes place inside cells, as seen in animals belonging to the *phylum Cnidaria*, is known as intracellular digestion.

F₁ (first filial generation): the offspring resulting from a genetic cross between *homozygous* parents.

■ If the F_1 generation are interbred, the F_2 (second filial generation) will be produced.

facilitated diffusion: the *diffusion* of a substance across a membrane, facilitated (assisted) by a protein *carrier molecule*.

■ Facilitated diffusion is a passive process (does not require energy) and will only take place if there is a diffusion gradient across the membrane. A molecule binds to a specific carrier protein, gets transported across the membrane and is released on the other side. Relatively large molecules, such as *glucose* or *amino acids*, diffuse much faster by facilitated diffusion than they would in the absence of the carrier proteins.

fatty acid: *organic* compounds that are usually found as constituents of *lipids*.

■ The general formula of a fatty acid is $CH_3(CH_2)_nCOOH$, consisting of a methyl (CH_3) group, a hydrocarbon chain of varying length and an acidic carboxyl (COOH) group. If the hydrocarbon chain does not contain any double bonds between carbon atoms, it is known as a saturated fatty acid, such as stearic acid. If double bonds are present, it is known as an unsaturated fatty acid, such as linoleic acid. The diagram below shows the structure of stearic acid and linoleic acid.

Stearic acid
($C_{17}H_{35}COOH$)
$CH_3-CH_2-CH_2-CH_2-CH_2-CH_2-CH_2-CH_2-CH_2-CH_2-CH_2-CH_2-CH_2-CH_2-CH_2-CH_2-CH_2-C\overset{O}{\underset{OH}{}}$

Linoleic acid
($C_{17}H_{31}COOH$)
$CH_3-CH_2-CH_2-CH_2-CH_2-CH=CH-CH_2-CH=CH-CH_2-CH_2-CH_2-CH_2-CH_2-CH_2-CH_2-C\overset{O}{\underset{OH}{}}$

female: an organism that produces female *gametes* (sex cells).

■ Female gametes are generally larger than male gametes and are produced in much smaller numbers. They are also immobile, meaning that male gametes have to be transported in them in order that *fertilisation* can occur. In mammals, the female gametes are ova (eggs) and the male gametes are *sperm*.

fermentation: the breakdown of *organic* molecules in the absence of oxygen.

■ Fermentation is a form of *anaerobic respiration*, such as the production of lactic acid by muscles when oxygen is in short supply. Fermentation in micro-organisms has been used by biotechnologists to produce a number of useful products, such as beer, bread, yoghurt, cheese and various *antibiotics*.

fertilisation: the fusion of male and female *gametes* during the process of *sexual reproduction* to produce a *zygote*.

■ Each gamete is *haploid* (one set of *chromosomes*) and so the zygote produced by fertilisation is *diploid* (two sets of chromosomes). In mammals, fertilisation involves the fusion of a *sperm* (male) and an *ovum* (female). In flowering plants, a male gamete nucleus fuses with an egg nucleus, following *pollination*.

■ *TIP* In flowering plants, fertilisation is not the same as pollination, which is simply the transfer of pollen from the *stamen* to the *stigma*. Similarly, in mammals, copulation (the depositing of sperm into the female reproductive tract) is not the same as fertilisation.

fertiliser: a substance that is added to soil in order to increase its productivity.

■ Fertilisers may be natural, such as composts or manure, or synthetic chemicals, such as nitrates and phosphates. Fertilisers can significantly increase crop yields, but if leached from the soil by rain, they may result in *eutrophication* of rivers and lakes.

■ *TIP* Fertilisers are not the same as *pesticides*, which are chemicals used to kill pests.

fetal haemoglobin: the type of *haemoglobin* found in the blood of a *fetus*.

■ Fetal haemoglobin has a slightly different molecular structure to adult haemoglobin, giving it a higher affinity for oxygen. This is important in the *placenta*, where fetal blood is able to pick up oxygen from the mother's blood in a very efficient manner. Following birth, fetal haemoglobin is replaced by adult haemoglobin.

■ *TIP* The higher affinity for oxygen of fetal haemoglobin means that its *oxygen dissociation curve* lies to the left of that for adult haemoglobin.

fetus (foetus): the stage of development of a mammalian *embryo* when the main features of the adult form can be seen.

■ In humans, the fetal stage lasts from eight weeks after *fertilisation* to birth.

fibrin: an insoluble protein that is the foundation of a blood clot.

■ Fibrin is formed from soluble fibrinogen during the process of *blood clotting*.

Filicinophyta: the plant *phylum* that contains the ferns.

■ Filicinophytes share the following features:
  ● *alternation of generations*, with the *sporophyte* being dominant
  ● *spore*-bearing structures called sporangia on the underside of leaves
  ● true roots, stems, leaves and *vascular* tissue

first filial generation: see F_1.

flagellum: a long hair-like structure found on the surface of cells which is involved in locomotion.

- Flagella (plural) are up to 150 µm long and are found on the surface of *bacteria*, *protoctists* and *sperm* cells. The possession of a flagellum makes these cells motile.
- **TIP** Biology students often confuse flagella with *cilia*. Although they are similar in structure, cilia tend to be much shorter and there are usually large numbers present on each cell.

florigen: a hypothetical *plant growth substance* that may stimulate flowering in plants.

- The production of florigen is thought to be stimulated by *phytochrome*, a pigment found in leaves. If florigen is formed in leaves and flowers form at the shoot tips, it is presumably a chemical messenger that is transported through the plant. However, florigen has never been isolated and some biologists are not convinced that this growth substance exists.

flower: a structure that contains the sexual reproductive organs in certain plants.

- Flowers are found in plants of the *phylum Angiospermophyta*. They are very variable in form, ranging from the small green flowers found in wind-pollinated plants to large, brightly-coloured flowers in plants pollinated by insects or other animals. *Pollination* must occur if *sexual reproduction* is to be successful. In other words, pollen has to be transferred from the *anther* to the *stigma* of the same or a different flower.

fluid mosaic model: a theory that describes the structure of *cell membranes*.

- The fluid mosaic model states that cell membranes are composed of a *phospholipid* bilayer containing a number of proteins, as shown in the diagram below. The protein molecules in the membrane may act as *enzymes*, carrier proteins for *active transport* or *facilitated diffusion*, or as specific *receptors* for *hormones*.

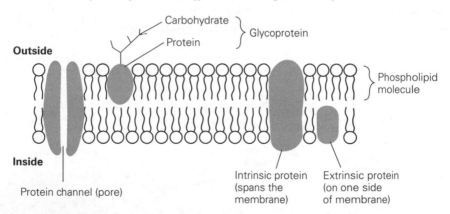

foetus: see *fetus*.

follicle-stimulating hormone (FSH): a mammalian *hormone* that stimulates the production of *gametes*.

- FSH is secreted by the anterior lobe of the *pituitary gland* and stimulates the production of ova (eggs) in females and *sperm* in males.
- **TIP** Make sure that you know how FSH interacts with *oestrogen, progesterone* and *luteinising hormone (LH)* to control the *menstrual cycle*.

food chain: a sequence of feeding relationships in an *environment*.

■ Food chains start with *producers*, which form the first *trophic level*. Producers are fed upon by primary *consumers*, which are then fed upon by secondary consumers, and so on. Food chains do not generally extend beyond a tertiary consumer, due to energy being lost from the system at each stage. Many animals feed at more than one trophic level and also compete with each other for the same food sources. Therefore, single food chains rarely exist and a *food web* is a more realistic representation of feeding relationships.

■ *e.g.* A food chain is shown in the diagram below.

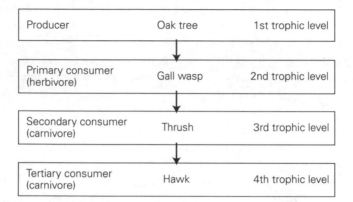

Producer	Oak tree	1st trophic level
Primary consumer (herbivore)	Gall wasp	2nd trophic level
Secondary consumer (carnivore)	Thrush	3rd trophic level
Tertiary consumer (carnivore)	Hawk	4th trophic level

■ *TIP* Remember that the arrows on a food chain represent the flow of energy. They always go from producers (1st trophic level) to consumers (higher trophic levels).

food web: a system of interconnected *food chains*.

■ *e.g.* A simplified food web is shown in the diagram below.

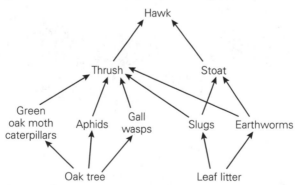

■ *TIP* Note that some animals can feed as both a primary *consumer* and a secondary consumer. Therefore it is sometimes difficult to define feeding relationships accurately.

fovea: part of the *retina* of the eye, containing a very large number of *cone cells*.

■ Images are usually focused on to the fovea, which is specialised for the perception of colour and fine detail.

f

fruit: the structure that encloses the *seeds* in a flowering plant.

■ A fruit develops from an *ovary* after *fertilisation* in the plant. Its function is to aid dispersal of seeds by animals or mechanical methods. In animal dispersal, edible fruits (e.g. cherries) are consumed by animals and the undigested seeds pass out in the faeces. Other fruits are sticky or have hooks so that they become temporarily attached to the fur or feathers of a passing mammal or bird. In mechanical dispersal, many fruits are light and often have hairs that increase their surface area and enable them to be blown by the wind (e.g. dandelions). Other fruits dry out and suddenly split, flinging the seeds over a wide area.

■ *TIP* Make sure that you know the roles of the various *plant growth substances* in the development of fruits.

FSH: see *follicle-stimulating hormone*.

Fungi: the *kingdom* containing mushrooms, toadstools, moulds and yeasts.

■ Fungi share the following features:
- they are *heterotrophic*
- they reproduce by means of *spores*
- they are usually made up of thread-like structures known as *hyphae*
- most have cell walls made of *chitin*

gall bladder: a small sac which stores *bile* produced by the *liver*.

■ The gall bladder is found just under the liver and it is connected to the small intestine by the bile duct. The release of bile is stimulated by the *hormone* cholecystokinin, when partly digested food from the stomach enters the small intestine.

gamete: a reproductive cell that fuses with another gamete to form a *zygote*.

■ Gametes are produced during *sexual reproduction* by *meiosis* and are *haploid* (contain a single set of *chromosomes*). The fusion of gametes is called *fertilisation* and the resulting zygote is *diploid* (having two sets of chromosomes). Female gametes are usually large, stationary and produced in small numbers, such as ova (eggs) in mammals. Male gametes are usually small, motile and produced in very large numbers, such as *sperm* in mammals.

gametogenesis: the formation of *gametes*.

■ In mammals, gametogenesis takes place in the *testes* in males (spermatogenesis) and the *ovaries* in females (oogenesis). Gamete-forming cells undergo *meiosis* and *differentiation* to form *sperm* or ova (eggs). This process is very similar in all organisms that reproduce sexually, resulting in the formation of large numbers of small gametes in males and small numbers of large gametes in females.

gametophyte: the stage in the life cycle of a plant in which *gametes* are produced.

■ The gametophyte stage alternates with the *sporophyte* stage (see *alternation of generations*). Gametophytes are *haploid* (contain one set of *chromosomes*) and therefore produce gametes by *mitosis*. The gametophyte may contain both male and female sex organs, as in *Filicinophyta* (ferns), or may exist in separate male and female forms, as in *Angiospermophyta* (flowering plants).

gastric juice: a mixture of hydrochloric acid, mucus and digestive *enzymes* secreted by gastric glands in the *stomach* wall.

■ The secretion of gastric juice is stimulated by the presence of food in the mouth and stomach. The most important enzymes in gastric juice are *pepsin*, which digests proteins into smaller polypeptides, and rennin, which coagulates milk protein. Both of these enzymes are secreted in an inactive form (pepsinogen and pro-rennin) and are activated by hydrochloric acid. The acid also provides the optimum *pH* for the action of these enzymes and helps to kill many pathogenic (disease-causing) *bacteria*.

gene: a section of *DNA* that codes for the production of a particular *polypeptide* or protein.

■ A gene may also be defined as a unit of heredity that is found at a specific position on a *chromosome*. Different forms of a particular gene are known as *alleles*.

■ *TIP* Many students confuse the terms gene and allele. Remember that a gene controls a character, such as eye colour, and that alleles control different forms of this character, such as blue, green or brown eyes.

gene mutation: any change in the *nucleotide* sequence of a *gene*.

■ Gene mutations are sometimes called point mutations because they occur at a single point on a *chromosome*. Nucleotide bases may be added (insertion), lost (deletion), changed (substitution) or rearranged (inversion). This may result in a change in one or more *amino acids* in the synthesised protein, which may affect its function.

■ *e.g.* Sickle-cell anaemia is caused by a substitution mutation that changes a single amino acid in *haemoglobin*. This results in distortion of *red blood cells*, reducing the amount of oxygen that can be carried. The effect of this mutation is summarised in the diagram below.

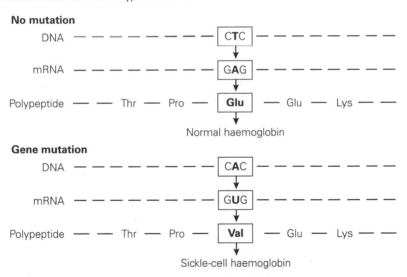

■ *TIP* Gene mutations do not always change the function of the protein produced. For example, the mutant sequence may code for the same amino acid; or the amino acid may change, but this does not affect the structure of function of the protein.

gene pool: all the *genes* and their different *alleles* that are present in a particular *population*.

■ The concept of a gene pool is useful when considering population genetics, as described by the *Hardy–Weinberg equilibrium*.

g

gene technology: the processes involved in changing the characteristics of an organism by inserting foreign *genes* into its *DNA*.

■ Gene technology is also known as genetic manipulation or genetic engineering. Some of the techniques used in gene technology are summarised in the flow-chart below.

Isolate the gene
- Locate the gene with a *DNA probe* and extract it with *restriction endonucleases*.
  or
- Produce the gene from a molecule of *messenger RNA (mRNA)* using reverse transcriptase.

Transfer the gene
- Cut open a *plasmid* with a restriction endonucleases, insert the gene and join the pieces of DNA together with DNA ligase.
- Insert the recombinant plasmid into a host *micro-organism*, e.g. a *bacterium*.
- Check that the bacterium has taken up the plasmid by testing for a marker gene on the plasmid, e.g. resistance to *antibiotics*.

Make the product
- Grow the bacteria containing the required gene in a fermenter. They multiply rapidly and produce large quantities of the desired product.

Gene technology has a wide range of practical applications, from the commercial production of *insulin* and vaccines, to the creation of transgenic animals and crop plants in agriculture.

gene therapy: the application of *gene technology* to alter or replace defective genes.

■ Gene therapy is currently being developed in the hope of curing a range of genetic diseases, such as cystic fibrosis.

generator potential: the rapid change in electrical charge across the cell surface membrane of a *receptor* as a result of a *stimulus*.

■ Stimulation of a receptor causes the electrical charge across the membrane to change from the *resting potential*. This change is called a generator potential. If the generator potential is sufficiently large (following a large stimulus), it will cause an *action potential* in the sensory *neurone* that is connected to the receptor.

genetic code: the method by which the genetic information in *DNA* controls the synthesis of specific proteins by the cell.

■ The sequence of *nucleotide* bases on a strand of DNA acts as the code for *protein synthesis*. Each triplet of bases on DNA translates into a *codon* on *messenger RNA (mRNA)*. Each codon represents an *amino acid* in the protein to be synthesised, or an instruction to start or stop protein synthesis.

■ *e.g.* AUG codes for 'start' in *translation* and for the amino acid methionine; GUG codes for the amino acid valine; UAG codes for 'stop' in translation.

genetic conservation: the preservation of genetic variation.

■ Modern agriculture depends upon a small number of specialised breeds which have little or no genetic variation. As a result, potentially useful *alleles* that may have been present in ancestral breeds will no longer be present. If environmental conditions change, farmers may wish to reintroduce these useful alleles into domesticated plants and animals. Therefore, it is necessary to conserve genetic material, either in seed banks (for plants) or by maintaining small populations of older breeds. Genetic conservation is also used to prevent the extinction of rare forms of wild plants and animals.

genetic counselling: the advice given to people from families with a history of genetic disease about the risks of having children.

■ Families with a history of diseases such as cystic fibrosis, haemophilia and sickle-cell anaemia are routinely offered genetic counselling. This may be accompanied by genetic screening, which involves testing to see whether an individual possesses an abnormal *gene* or *chromosome mutation*.

genetic fingerprinting: the technique in which an individual's *DNA* is analysed to reveal a unique pattern of *nucleotide* sequences.

■ Genetic fingerprinting is used for identification purposes in forensic science, paternity disputes and veterinary science. The techniques used are summarised in the flow chart below.

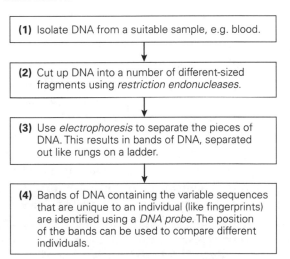

genotype: the genetic composition of an organism.

■ The genotype of an organism is the sum total of all the *alleles* present. However, we normally only consider one or two *genes*. Genotypes are usually represented by symbols, as shown in the example that follows.

■ *e.g.* The inheritance of height in garden peas is controlled by two alleles, T (the tall allele) and t (the short allele). The possible genotypes of a pea plant are therefore TT, Tt or tt, as there must be two alleles present for each characteristic. TT and tt are known as *homozygous* genotypes (the alleles are the same) and Tt is known as the *heterozygous* genotype (the alleles are different).

■ *TIP* Try not to confuse genotype with *phenotype*, which is the physical expression of the genotype (e.g. tall or short in pea plants) and may be influenced by the *environment*.

geotropism: the directional growth of part of a plant in response to gravity.

■ Roots are positively geotropic, because they grow towards gravity, while stems are negatively geotropic, because they grow away from gravity.

germination: the initial growth of a *seed*, using stored food reserves.

■ Germination begins when the seed starts to absorb water. The *plumule* (shoot) and *radicle* (root) emerge from the plant *embryo* and grow upwards and downwards respectively. Food reserves come from *endosperm* tissue or from the *cotyledons*. Germination is complete when the seedling begins to obtain its own nutrients by *photosynthesis*.

■ *TIP* Make sure that you know the roles of the various *plant growth substances* in the process of germination.

gibberellin: a *plant growth substance* which acts mainly as a growth stimulator.

■ Gibberellin stimulates *germination*, cell elongation in the stem and the breaking of dormancy in buds.

■ *TIP* Make sure that you know how gibberellin interacts with other plant growth substances, such as *abscisic acid*, *auxin*, *cytokinin* and *ethene*.

gill: the gas exchange organ of many aquatic animals, such as fish, mussels and tadpoles.

■ Gills are generally thin, permeable to respiratory gases, have a large surface area and a good blood supply. The diffusion of oxygen into the blood and carbon dioxide out, is particularly efficient in fish because the blood flows in an opposite direction to the water passing over the gills (see *counter-current system*).

gland: a group of cells that is specialised for secretion.

■ *e.g.* There are two types of gland in mammals. *Endocrine glands* secrete *hormones* directly into the blood (e.g. the *thyroid* gland). *Exocrine glands* secrete *enzymes* or other substances into a duct (e.g. the salivary glands).

■ *TIP* Note that the *pancreas* is both an endocrine and an exocrine gland.

glomerular filtrate: the fluid found in the *Bowman's capsule* of a kidney *nephron*.

■ Glomerular filtrate is formed by *ultrafiltration* from the *glomerulus*. It has the same composition as blood *plasma*, except that it does not contain any of the larger components, such as *plasma proteins* or cells.

■ *TIP* Note that glomerular filtrate contains many useful substances filtered out of the blood, such as *glucose*. These substances have to be reabsorbed from the nephron back into the blood. Glomerular filtrate is therefore not the same as urine.

glomerulus: a group of *capillaries* enclosed by the cup-shaped end of a kidney *nephron*.

- Blood enters the glomerular capillaries at high pressure. This forces *glomerular filtrate* into the *Bowman's capsule* and down the nephron, a process known as *ultrafiltration*.

glucagon: a *hormone* involved in the control of blood *glucose*.

- Glucagon is synthesised by the α-cells in the islets of Langerhans in the *pancreas*. It is secreted in response to a decreased concentration of glucose in the blood and brings about an increase in the *blood glucose level*, as shown in the diagram below.

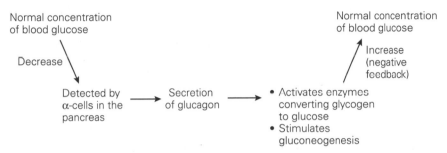

- **TIP** Try not to confuse glucagon with *glycogen*, which is a storage carbohydrate found in animals. The words look very similar, so it is vital that you use the correct spelling. The action of glucagon is often confused with that of *insulin*, which has the opposite effects. Insulin is secreted in response to an increased concentration of glucose in the blood, and brings about a decrease in the blood glucose level. In other words, insulin and glucagon are *antagonistic*.

gluconeogenesis: the synthesis of *glucose* from non-carbohydrate sources, such as *lipid* and *protein*.

- Gluconeogenesis occurs when the *blood glucose level* is low and *glycogen* stores in the liver have been depleted. The synthesis of glucose is essentially a reversal of *glycolysis* and it can use many sources, such as *amino acids* and glycerol.

glucose: a hexose (six-carbon) *monosaccharide*.

- Glucose has the molecular formula $C_6H_{12}O_6$. It is the main *substrate* for *respiration* and long chains of glucose make up *cellulose, starch* and *glycogen*. There are two forms (isomers) of glucose, as shown in the diagram below.

glycogen: a storage carbohydrate found in animals.

■ Glycogen is a *polysaccharide*, made up of long, branched chains of α-*glucose*. In mammals, it is mainly stored in liver and muscle cells as a source of glucose for use in *respiration*.

■ *TIP* Try not to confuse glycogen with *glucagon*, which is a *hormone* involved in the control of blood glucose. The words look very similar, so it is vital that you use the correct spelling.

glycogenesis: the conversion of *glucose* to *glycogen*.

■ Glycogenesis is stimulated by *insulin* and takes place in liver and muscle cells. This results in a reduction in *blood glucose levels*.

■ *TIP* Try not to confuse glycogenesis with *glycogenolysis*, which is the conversion of glycogen to glucose (the opposite of glycogenesis). Remember that 'genesis' means 'to create' and 'lysis' means 'to break down'.

glycogenolysis: the conversion of *glycogen* to *glucose*.

■ Glycogenolysis is stimulated by *glucagon* and *adrenaline*, and takes place in liver and muscle cells. This results in an increase in *blood glucose levels*.

■ *TIP* Try not to confuse glycogenolysis with *glycogenesis*, which is the conversion of glucose to glycogen (the opposite of glycogenolysis). Remember that 'genesis' means 'to create' and 'lysis' means 'to break down'.

glycolysis: the first stage of *respiration*, in which *glucose* is broken down to form *pyruvate*.

■ Glycolysis occurs in the *cytoplasm* of cells and is common to both *aerobic respiration* and *anaerobic respiration*. The biochemical reactions that take place during glycolysis are summarised in the diagram below.

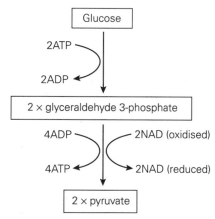

i.e. glucose + 2NAD (oxidised) + 2ADP + 2Pi $\longrightarrow$ 2 × pyruvate + 2NAD (reduced) + 2ATP

glycoprotein: a molecule composed of a *carbohydrate* linked to a protein.

■ Glycoproteins are important components of *cell membranes*, where they function as *receptors* for *hormones*, or as *antigens*. They are also found in body fluids, such as *mucus*.

Golgi apparatus: a collection of membranes and *vesicles* found in the *cytoplasm* of cells.

g

- The Golgi apparatus is an *organelle* with a number of functions, including the synthesis of *glycoproteins*; the *secretion* of *enzymes* and *hormones*; and the formation of *lysosomes*.
- *TIP* Cells which are involved in the secretion of substances contain large amounts of Golgi apparatus.

Gram staining: a staining method used to classify *bacteria*.

- Bacteria are stained first with crystal violet, then with a red stain. Gram-positive bacteria have thick *cell walls* and retain the crystal violet stain, appearing dark blue under a microscope. Gram-negative bacteria have thin cell walls and do not retain the crystal violet stain, appearing red under a microscope.

greenhouse effect: the way in which heat is trapped in the earth's atmosphere.

- Radiation from the sun warms the earth's surface, which then radiates heat into the environment. Carbon dioxide prevents some of this heat from leaving the atmosphere and so the air temperature rises. It is important to note that the greenhouse effect is useful, otherwise the earth would be too cold to support life. However, human activities have caused an increase in the amount of atmospheric carbon dioxide (and other 'greenhouse gases', such as methane and nitrogen oxides), due to deforestation and the combustion of fossil fuels. As more and more heat is retained in the atmosphere, this will lead to an increase in the average temperature of the earth, a phenomenon known as global warming.
- *TIP* Remember that *chlorofluorocarbons (CFCs)* are not directly responsible for the greenhouse effect.

growth: a permanent increase in size of an organism.

- Growth can be measured by an increase in cell size or cell numbers, or an increase in volume or mass. It may continue throughout the life of the organism, as seen in trees, or it may cease at maturity, as in humans.

guanine: a *purine* nitrogenous base that is found in *DNA* and *RNA*.

- In DNA, guanine always forms a bond with *cytosine*, a *pyrimidine* nitrogenous base.

habitat: the particular *environment* in which an organism lives.

■ A habitat is usually a defined area with characteristic *abiotic* (non-living) and *biotic* (living) conditions.

■ *e.g.* A pond, a rocky shore, a hedgerow and a woodland are all habitats containing characteristic *communities* of living organisms.

haem: an iron-containing *organic* molecule.

■ Haem is usually found bound to a protein, as seen in *haemoglobin, myoglobin* and *cytochromes.*

haemoglobin: an iron-containing protein found in *red blood cells.*

■ Haemoglobin consists of four *polypeptide* chains, each linked to a *haem* group. Its function is to transport oxygen around the body of many animals, each haem group combining reversibly with an oxygen molecule. This can be summarised as follows:

$$Hb + 4O_2 \underset{\substack{\text{respiring} \\ \text{tissues}}}{\overset{\text{lungs}}{\rightleftarrows}} Hb(O_2)_4$$

haemoglobin   oxygen   oxyhaemoglobin

The relationship between the concentration of oxygen in the blood and the percentage saturation of haemoglobin is shown by the *oxygen dissociation curve.* In addition to its oxygen-carrying function, haemoglobin also transports some carbon dioxide (as carbamino compounds, not carboxyhaemoglobin) and is an important *buffer* in the blood.

haploid: any cell in which the *nucleus* contains one set of *chromosomes.*

■ Most body cells in humans are *diploid* (sometimes referred to as 2n), possessing a full complement of 46 chromosomes in the nucleus. *Gametes*, however, are haploid as they have only half the number of chromosomes that are present in a body cell. For example, a human sperm cell contains 23 chromosomes, which is known as the haploid (n) number.

■ *TIP* Remember that the haploid number of chromosomes is variable, and differs from one species to another. For example, the haploid number in pea plants is 7.

Hardy–Weinberg equilibrium: states that, provided certain conditions exist, the frequency of *alleles* in a *population* will remain constant.

■ For the Hardy–Weinberg equilibrium to exist, the following conditions must be satisfied:

- the population is large
- there is no *natural selection*
- *mutations* do not occur
- mating is random
- there is no migration into or out of the population

In practice, one or more of these conditions is often not satisfied. In other words, the frequency of alleles in a population does sometimes change and *evolution* may take place.

HCG: see *human chorionic gonadotrophin*.

heart: a muscular organ that pumps *blood* around the body of an animal.

■ The generalised structure of a mammalian heart is shown in the diagram below.

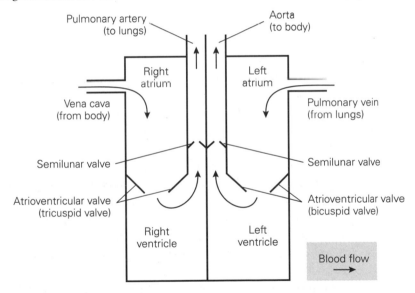

The pumping action of the heart is described by the *cardiac cycle*.

■ *TIP* It is easy to confuse the left and right sides of the heart, the names of the associated blood vessels and the direction of blood flow. It is important that you know and understand this information.

hepatic portal vein: a vessel that transports blood from the small intestine to the liver.

■ Blood carried by the hepatic portal vein carries all the soluble products of *digestion* that have been *absorbed* in the small intestine.

herbicide: a chemical substance that kills plants.

■ Herbicides are manufactured to kill weeds and other unwanted plants. Non-selective herbicides (e.g. paraquat) kill a wide variety of plants and are used mainly for land clearance. Selective herbicides (e.g. 2,4-D) kill certain *species* only, so can be used on lawns and crop plants. The application of herbicides may disrupt *ecosystems* by destroying *habitats*, interfering with *food webs*, or killing non-target organisms.

herbivore: an animal that feeds on plants.

■ In a *food web*, a herbivore is a primary *consumer* and occupies the second *trophic level*.

■ *e.g.* Sheep, rabbits and locusts are herbivores.

heterotrophic nutrition: a type of nutrition in which an organism gains its required nutrients by consuming complex *organic* molecules.

■ All animals, fungi and most *bacteria* are heterotrophic. There are three types of heterotrophic nutrition.

● Holozoic nutrition — feeding on organic matter obtained from the bodies of other organisms. Food is digested internally by *enzymes* and the soluble products of digestion are then absorbed.

● Saprophytic nutrition — feeding on organic matter from dead and decaying organisms. Food is digested externally by enzymes and the soluble products of digestion are then absorbed.

● Parasitic nutrition — feeding on organic matter in the living tissues of a host organism. Digestion is not normally required as soluble nutrients can be absorbed directly from the host.

■ *TIP* Do not confuse heterotrophic nutrition with *autotrophic nutrition*, which is the synthesis of organic molecules from simple inorganic molecules, such as carbon dioxide and water.

heterozygous: a condition in which the *alleles* of a particular *gene* are different.

■ *e.g.* The height of pea plants is determined by a gene with two alleles. The allele for a tall plant (T) is *dominant* to that for a short plant (t). The *genotype* of a heterozygous plant will be Tt and its *phenotype* (appearance) will be tall.

hexose: a *monosaccharide* containing six carbon atoms.

■ *e.g. Glucose*, fructose and galactose are all hexoses with the molecular formula $C_6H_{12}O_6$.

histocompatibility: the extent to which a tissue from one organism will be tolerated by the immune system of another organism.

■ In general, different organisms have different histocompatibility *antigens* on the surface of their cells. When tissues are transplanted, the graft is usually recognised as foreign because of these antigens. As a result, the graft is rejected by the host's immune system. This may not happen if the donor and recipient are closely related, as histocompatibility antigens are determined genetically. In these cases, there is a much greater chance that the transplant will be successful.

histone: a protein that is found as a major constituent of *chromosomes*.

■ Chromosomes consist of *DNA* coiled around histones and other proteins.

HIV: see *human immunodeficiency virus*.

homeostasis: the maintenance of a constant internal environment.

■ Cells require certain conditions to function efficiently. Intercellular fluid (and therefore blood) must contain oxygen and nutrients, and have the optimum *pH*, temperature and *water potential*. Homeostasis represents the set of physiological mechanisms involved in regulating this internal environment. Most homeostatic mechanisms are coordinated by *hormones* and involve *negative feedback*, as shown in the diagram below.

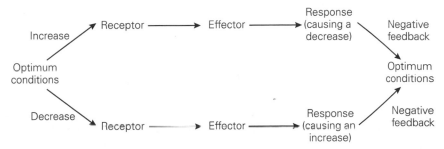

■ *e.g. Thermoregulation*, the control of respiratory gases in the blood, *osmoregulation* and the regulation of *blood glucose*.

hominid: any member of the primate family Hominidae, which includes modern humans and their recent ancestors.

■ Hominids are able to walk upright and they have relatively large brains. These features have allowed other characteristics to develop, such as the use of tools, as described by *cultural evolution*. The diagram below shows how modern humans (**Homo sapiens**) may have evolved from an ancestral group (**Australopithecus**).

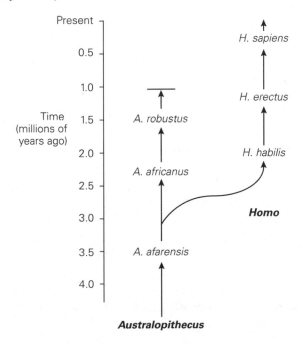

Homo: the genus containing modern humans and their recent ancestors.

■ The oldest **Homo** fossils are those of **Homo habilis**, from about 1.75 million years ago. **Homo habilis** was about 1–1.5 m tall with more human-like features and a larger brain than **Australopithecus**. **Homo erectus** emerged about 1.5 million years ago and looked like modern humans, except for the prominent ridge above the eyes and a reduced forehead and chin. **Homo erectus** had basic stone tools and used fire. **Homo sapiens** (modern humans) appeared about 100 000 years ago.

homologous: features of organisms that have the same evolutionary origin but have developed different functions.

■ *e.g.* The wings of a bat, the flippers of a whale and the legs of a dog are homologous organs because they all evolved from the limbs of a common ancestor.

■ *TIP* Structures with similar functions, but a different evolutionary origin, are called analogous features, e.g. the wings of a bird and the wings of an insect.

homologous chromosomes: a pair of *chromosomes* with similar structural features.

■ In a *diploid* cell, the nucleus contains two of each type of chromosome. Human body cells, for example, contain 23 pairs of chromosomes. One of each pair is the maternal chromosome and the other is the paternal chromosome. These chromosomes are the same length and contain the same *genes* (although perhaps different *alleles*), arranged in the same sequence.

homozygous: a condition in which the *alleles* of a particular *gene* are the same.

■ *e.g.* The height of pea plants is determined by a gene with two alleles. The allele for a tall plant (T) is *dominant* to that for a short plant (t). The *genotype* of a homozygous plant will be either TT or tt, and its *phenotype* (appearance) will be tall (TT) or short (tt).

hormone: a chemical messenger that is transported in the blood.

■ Hormones are secreted into the blood by *endocrine glands*. They are transported to target cells, where they bind to specific *receptors* on the cell surface membrane. The hormone then influences the *physiology* of the cell in some way, by changing membrane permeability, altering the activity of *enzymes*, or activating or inhibiting *genes*.

■ *e.g.* Insulin is a hormone secreted by the *pancreas* in response to an increased *blood glucose level*. It stimulates liver and muscle cells to take up *glucose* from the blood and convert it to *glycogen*, thus lowering the blood glucose level.

human chorionic gonadotrophin (HCG): a *hormone* produced early in pregnancy by the *placenta* of mammals.

■ The main function of HCG is to maintain the *corpus luteum* in the *ovary*. This enables the corpus luteum to continue to produce *progesterone*, thereby maintaining the lining of the *uterus* for the developing *embryo*. Pregnancy can be diagnosed very early by testing for the presence of HCG in urine.

human immunodeficiency virus (HIV): a *virus* that causes acquired immune deficiency syndrome (AIDS) in humans.

■ HIV is transmitted in blood *plasma* and sexual fluids. The virus binds to specific proteins on the surface of a type of *white blood cell* called a T-helper cell, and inserts its *DNA* into the human DNA. After a period of time, new viruses are synthesised and released from the T-helper cell, which is destroyed. A reduction in the number of these cells in the body weakens the immune response, a condition known as AIDS. The infected individual will be susceptible to attack by pathogenic microorganisms, such as the *bacteria* that cause pneumonia. It is these secondary infections that cause the deaths associated with AIDS.

hydrogen bond: a weak electrostatic attraction between a hydrogen atom and an oxygen atom.

■ In certain molecules, hydrogen atoms tend to have a slightly positive charge and oxygen atoms tend to have a slightly negative charge, meaning that hydrogen bonding can take place.

■ *e.g.* Hydrogen bonds occur between water molecules, as shown in the diagram below.

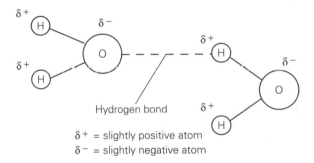

$\delta +$ = slightly positive atom
$\delta -$ = slightly negative atom

Hydrogen bonds also play an important role in some large biological molecules, such as *cellulose* and *DNA*.

hydrolase: an *enzyme* that catalyses the addition of water to a molecule.

■ The breakdown of a molecule by the addition of water is known as *hydrolysis*. Most of the enzymes involved in *digestion* are hydrolases.

■ *e.g.* *Lipase* hydrolyses *lipids* into *fatty acids* and glycerol.

hydrolysis: the breakdown of a molecule by the addition of water.

■ Many biological molecules are broken down by hydrolysis during *digestion*. The bonds between the sub-units of the large food molecules are broken by adding water.

■ *e.g.* The hydrolysis of a *lipid* yields *fatty acids* and glycerol.

■ *TIP* The reverse of hydrolysis is *condensation*, the construction of large molecules by the removal of water.

hydrostatic skeleton: the supportive internal framework of soft-bodied invertebrates.

■ Hydrostatic skeletons are based on a fluid-filled body cavity. Fluids are incompressible and so this internal framework can be used for support and locomotion.

■ *e.g.* Earthworms have a hydrostatic skeleton based on a fluid-filled *coelom* enclosed by a muscular body wall.

hyperglycaemia: a condition characterised by abnormally high levels of *glucose* in the blood.

■ Hyperglycaemia occurs in people suffering from *diabetes*, due to insufficient production of *insulin*. Symptoms of the condition include dehydration, weight loss and glucose in the urine. It can be treated by injections of insulin.

■ *TIP* Try not to confuse this term with *hypoglycaemia*, which is an abnormally low level of glucose in the blood. Remember that 'hyper' means 'over' or 'more than', e.g. a hyperactive person is too active; 'hypo' means 'under' or 'less than', e.g. a hypodermic syringe is designed to go under the skin (dermis).

hypertonic: a solution that has a lower (more negative) *water potential* than another solution.

■ Solution A will be hypertonic to solution B if it contains less water or a higher concentration of solutes. If the two solutions were separated by a *partially permeable membrane*, we would expect water to move from B to A by *osmosis*.

■ *TIP* The opposite of hypertonic is hypotonic, which refers to a solution that has a higher (less negative) water potential than another solution.

hyphae: a mass of thread-like structures that make up the body tissue of *Fungi*.

■ Each hypha (singular) is surrounded by a *cell wall* which is usually made of *chitin*. A mass of hyphae is sometimes referred to as a *mycelium*.

hypoglycaemic: a condition characterised by abnormally low levels of *glucose* in the blood.

■ Hypoglycaemia occurs in people suffering from *diabetes*, due to insufficient intake of glucose or an excess of injected *insulin*. Symptoms of the condition include weakness, dizziness, sweating and, if untreated, coma. It can be treated by consuming glucose.

■ *TIP* Try not to confuse this term with *hyperglycaemia*, which is an abnormally high level of glucose in the blood. Remember that 'hyper' means 'over' or 'more than', e.g. a hyperactive person is too active; 'hypo' means 'under' or 'less than', e.g. a hypodermic syringe is designed to go under the skin (dermis).

hypothalamus: part of the *brain* which plays an important role in *homeostasis*.

■ The hypothalamus is the main coordinating centre for the *autonomic nervous system* and it controls a number of functions, such as feeding, drinking, sleeping, *thermoregulation* and *osmoregulation*. Many of its effects are achieved via the *pituitary gland*. For example, the hypothalamus secretes *thyrotrophin-releasing hormone* (*TRH*) in response to low levels of *thyroxine* in the blood. TRH stimulates the pituitary gland to produce *thyroid-stimulating hormone* (*TSH*), which promotes the synthesis and secretion of thyroxine, thus increasing its concentration in the blood.

ileum: part of the small intestine in which the process of *digestion* is completed and *absorption* takes place.

■ Cells in the wall of the ileum secrete a number of digestive *enzymes,* including sucrase (which digests *sucrose* to *glucose* and fructose) and dipeptidase (which digests dipeptides to *amino acids*). The ileum is adapted for absorption in a number of ways:

- it is very long and highly folded, providing a large surface area
- the surface area is further increased by the presence of numerous villi (singular, *villus)*, which have an *epithelium* covered in microvilli
- the villi have a large number of *capillaries* and contain lacteals (part of the *lymph* system), so the absorbed products can be quickly transported away

Monosaccharides and *amino acids* are absorbed into capillaries by *diffusion* and *active transport*; *fatty acids* and glycerol are absorbed into the lacteals.

immobilisation: a technique used in *biotechnology* in which *enzymes* are fixed to an unreactive material.

■ Enzymes are usually immobilised by fixing them to beads, which are then placed in a column. The *substrate* is converted into products as it passes down the column. There are several advantages to immobilisation. The process is cheaper because the enzyme can be easily recovered at the end of the reaction and used again. It also helps avoid contamination of the products and can make certain enzymes tolerant of higher temperatures and a wider *pH* range.

■ *TIP* Immobilisation does not involve stopping an enzyme from working.

immune: a term used to describe an animal that is not susceptible to infection by certain disease-causing organisms, or that remains unaffected by their toxins.

■ Immunity depends on the presence in the blood of *antibodies* and *white blood cells*, which produce an immune response when the body is infected by pathogenic (disease-causing) organisms. Immunity may be inherited or acquired during the lifetime of the animal by the transfer of antibodies, exposure to infection, or vaccination.

immunoglobulin: see *antibody*.

implantation: the process by which an *embryo* becomes embedded in the lining of the *uterus* of a mammal.

■ Following implantation, the embryo continues to develop and the *placenta* is formed.

impulse: a signal that travels along a *neurone* (nerve cell).

■ Information is transmitted through the nervous system by impulses. The generation of an impulse, and its transmission along a neurone, is described in the diagram below.

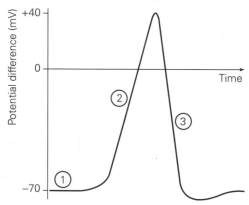

(1) Resting potential: a high concentration of sodium ions (Na^+) outside the cell. The inside is negative relative to the outside.

(2) Depolarisation: stimulation of the nerve cell causes sodium ions to enter the cell, making it positive relative to the outside.

(3) Repolarisation: potassium ions (K^+) leave the cell, restoring the negative potential difference. A sodium–potassium pump then restores the true resting potential by actively transporting potassium into the cell and sodium out.

■ *TIP* Students often have difficulty when trying to describe the nature of a nerve impulse in examinations. Make sure that you learn this information thoroughly. Remember that nerves carry impulses and not messages.

induced fit hypothesis: a model that describes the action of an *enzyme*.

■ The induced fit hypothesis proposes that the interaction of an enzyme with its *substrate* causes the *active site* of the enzyme to change shape slightly in order to allow the formation of an enzyme–substrate complex. In other words, the substrate induces the enzyme to fit. This model is generally preferred to the *lock and key hypothesis*, which argues that the active site has a rigid (unchanging) shape, into which the substrate fits exactly.

inorganic: any compound that is not based on carbon.

■ Inorganic compounds do not contain carbon, with the exception of carbon dioxide and carbonate salts. Many inorganic *ions* are important biologically, such as sodium (Na^+), *calcium* (Ca^{2+}) and phosphate (PO_4^{3-}).

■ *TIP* All compounds can be classified as either inorganic or *organic* (based on carbon).

insulin: a *hormone* involved in the control of blood *glucose*.

■ Insulin is synthesised by the β-cells in the *islets of Langerhans* in the *pancreas*.

It is secreted in response to an increased concentration of glucose in the blood and brings about a decrease in the *blood glucose level*, as shown in the diagram below.

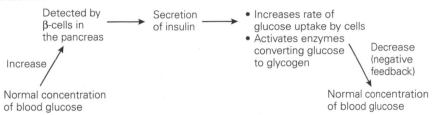

■ *TIP* The action of insulin is often confused with that of *glucagon*, which has the opposite effects. Glucagon is secreted in response to a decreased concentration of glucose in the blood, and brings about an increase in the blood glucose level. In other words, insulin and glucagon are *antagonistic*.

interferon: a protein molecule that prevents the replication of *viruses*.

■ Interferon may also protect the body against some forms of cancer. These useful properties have led to commercial production of interferon by *biotechnology*.

interspecific competition: competition for resources that occurs between members of different *species*.

■ Interspecific competition occurs mainly between organisms on the same *trophic level* competing for the same food supply.

■ *TIP* Try not to confuse this term with *intraspecific competition*, which refers to competition for resources that occurs between members of the same species. Remember that 'inter' means between different things and 'intra' means within something.

intraspecific competition: competition for resources that occurs between members of the same *species*.

■ Members of the same species may compete for food, mates or nesting sites.

■ *TIP* Try not to confuse this term with *interspecific competition*, which refers to competition for resources between members of different species. Remember that 'inter' means between different things and 'intra' means within something.

in vitro: any biological process that occurs outside the body, in an artificial situation.

■ in vitro means 'in glass' or 'in a test-tube'. The term usually refers to the observation of a biological process occurring in a laboratory setting.

■ *e.g.* in vitro *fertilisation* involves the external fertilisation of a human *ovum* (egg) and the subsequent *implantation* of the *embryo*.

■ *TIP* Try not to confuse this term with *in vivo*, which refers to biological processes that take place within living organisms.

in vivo: any biological process that occurs within a living organism.

■ in vivo usually refers to the observation of biological processes that take place in the natural environment.

■ *TIP* Try not to confuse this term with *in vitro*, which refers to biological processes that take place outside the body.

iodine: a nutrient that is required in very small quantities by living organisms.

■ Iodine is an important constituent of *thyroxine*, a *hormone* secreted by the *thyroid gland* which helps to control *basal metabolic rate*. A shortage of iodine in the diet leads to enlargement of the thyroid gland, a condition known as goitre.

ion: an electrically charged *atom* or *molecule*.

■ Atoms or molecules may either gain *electrons* and become negatively charged (anions), or lose electrons and become positively charged (cations).

■ *e.g.* Hydrogen ions (H^+) and hydrogencarbonate ions (HCO_3^-).

ionising radiation: radiation that is able to remove *electrons* from *atoms* or *molecules*.

■ Ionising radiation can be very damaging to body tissues and may increase the risk of *mutation* and cancer.

■ *e.g.* Ultraviolet light, X-rays and radioactivity.

iron: a nutrient that is required in small quantities by living organisms.

■ Iron is an important constituent of *cytochromes* and *haemoglobin*. It also activates catalase (an intracellular *enzyme* that breaks down hydrogen peroxide) and is required for the synthesis of *chlorophyll*. A lack of iron in the diet of humans leads to anaemia.

islets of Langerhans: small groups of cells found in the *pancreas*.

■ The α-cells of the islets of Langerhans secrete *glucagon* and the β-cells secrete *insulin*. These cells therefore function as an *endocrine gland* and play an important role in regulating the *blood glucose level*.

isolation: the process by which individuals of the same *species* are prevented from reproducing.

■ Individuals may be prevented from reproducing due to a physical barrier preventing interaction, such as a mountain range or river. This is known as geographical isolation and may lead to *allopatric speciation*. Alternatively, reproduction may be prevented by various other mechanisms associated with mating, *fertilisation* and the viability of offspring, known collectively as reproductive isolation. This may lead to *sympatric speciation*.

isomer: a *molecule* that contains the same *atoms* as another molecule, but has a different structure.

■ *e.g.* α-glucose and β-glucose are isomers as they have the same formula ($C_6H_{12}O_6$), but slightly different structures, as shown in the diagram below.

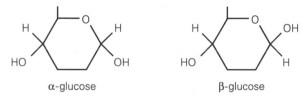

α-glucose   β-glucose

isotonic: a solution that has the same *water potential* as another solution.

■ Solution A will be isotonic to solution B if it has the same water content or contains an identical concentration of solutes. If the two solutions were separated by a partially permeable membrane, we would expect no net movement of water between A and B by *osmosis*.

■ *TIP* Remember that 'iso' means 'the same', so *isomers* have the same formula and isotonic solutions have the same water content.

isotope: an *atom* of the same *element* as another atom, but having a different mass.

■ All atoms of a given element contain the same number of protons in their nucleus, but may have different numbers of neutrons. This means that there may be several isotopes of the same element, some of which may be radioactive.

■ *e.g.* Nitrogen atoms usually have a mass of 14 ('normal' nitrogen), but may also have a mass of 15 ('heavy' nitrogen).

■ *TIP* Isotopes can be used to see what happens to particular atoms during biological processes. For example, Meselson and Stahl used isotopes of nitrogen to provide evidence for the *semi-conservative replication* of *DNA*.

kidneys: a pair of organs found in mammals which play an important role in *excretion* and *osmoregulation*.

■ Each kidney is composed of millions of *nephrons*, which are the structural and functional units of the organs. The nephrons are responsible for *ultrafiltration*, the selective reabsorption of useful products and the excretion of nitrogenous waste in the form of urine. They also regulate the removal of water from the body and are therefore important in osmoregulation.

kingdom: the largest group used in *classification*.

■ All living organisms can be classified into one of five kingdoms: *Prokaryotae, Protoctista, Fungi, Plantae* or *Animalia*.

Krebs cycle: a cyclical series of biochemical reactions which occur during *aerobic respiration*.

■ The Krebs cycle takes place in the matrix of *mitochondria* and the reactions can be summarised by the diagram below.

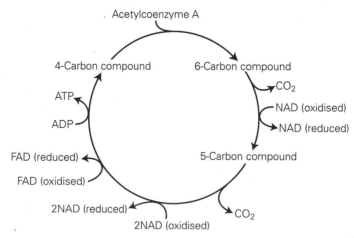

i.e. acetyl (2C) + 2NAD (oxidised) + 2FAD (oxidised) + ADP + Pi

$\longrightarrow$ 2CO$_2$ + 2NAD (reduced) + FAD (reduced) + ATP

■ *TIP* The Krebs cycle is also known as the citric acid cycle or the tricarboxylic acid (TCA) cycle.

lactation: the production of milk by mammary glands.

▨ Lactation occurs after birth of the young and is a two-stage process.
- Secretion of milk: stimulated by the hormone *prolactin*.
- Release of milk: stimulated by the hormone oxytocin, which is released in response to the sucking action of the infant.

lactose: a *disaccharide* found in milk.

▨ Lactose is composed of a *glucose* molecule linked to a galactose molecule.

leguminous plant: a plant that plays an important role in the *nitrogen cycle*.

▨ Leguminous plants are *dicotyledons* which can form a *mutualistic* relationship with nitrogen-fixing *bacteria* (*Rhizobium*) found in root nodules. The bacteria convert nitrogen in the air to ammonium compounds, some of which enter the soil. As a result, leguminous plants improve soil fertility and are used in crop rotations.

▨ *e.g.* Clover, peas and beans.

lens: a transparent structure in the eye which is responsible for focusing light onto the *retina*.

▨ The lens is a flexible structure located behind the iris and attached by suspensory ligaments to the ciliary body. The process of changing the shape of the lens in order to focus on near or far objects is called *accommodation*.

lenticels: small pores found in the stems of woody plants.

▨ Lenticels are made of cork and function as sites of gas exchange between the atmosphere and internal tissues.

▨ *TIP* Try not to confuse lenticels with *stomata*, which also function as sites of gas exchange, but can open or close and are found mainly on the underside of leaves.

LH: see *luteinising hormone*.

ligament: a tough but flexible *connective tissue* that holds bones together at a movable joint.

▨ Ligaments are composed mainly of *collagen* and are important in preventing dislocation.

▨ *TIP* Students often confuse ligaments with *tendons*, which are bands of connective tissue that attach muscle to *bone*.

L

ligase: any of a group of *enzymes* that are important in the synthesis and repair of many biological molecules.

■ *e.g.* DNA ligase is an enzyme that is involved in the synthesis of *DNA* and is also used in *gene technology*.

light-dependent reaction: the process in which light energy is absorbed by *chlorophyll* and used to produce *ATP* and reduced *NADP* during *photosynthesis*.

■ The light-dependent reaction consists of two processes:

- cyclic *photophosphorylation*, in which light energy causes the emission of a high-energy *electron* from chlorophyll. This electron is passed back to the chlorophyll molecule via an *electron transport chain*, generating ATP.

- non-cyclic photophosphorylation, in which light energy causes the emission of a high energy electron from chlorophyll. This electron is passed along an electron transport chain (generating ATP) and is used to reduce NADP. Chlorophyll has its electron replaced by the *photolysis* of water, resulting in the production of oxygen.

The light-dependent reaction is summarised in the diagram below.

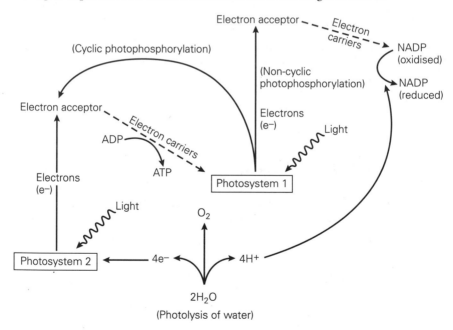

(Photolysis of water)

■ *TIP* Try not to confuse this process with the *light-independent reaction*, which does not require light and involves the reduction of carbon dioxide to *carbohydrate* and other *organic* molecules.

light-independent reaction: the process in which carbon dioxide is fixed and reduced to *carbohydrate* during *photosynthesis*.

■ Carbon dioxide is fixed by *ribulose bisphosphate (RuBP)*, generating two molecules of glycerate 3-phosphate (phosphoglyceric acid). *ATP* and reduced *NADP* (from the *light-dependent reaction*) then convert glycerate 3-phosphate to glyceraldehyde 3-phosphate (triose phosphate). This can be used to regenerate RuBP (to fix more carbon dioxide), or built up into carbohydrates, *lipids, proteins* and *nucleic acids*. The light-independent reaction is summarised in the diagram below.

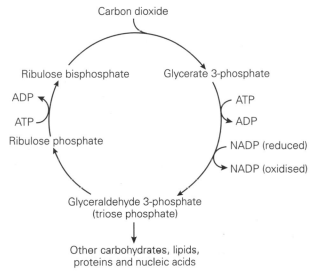

Carbon dioxide

Ribulose bisphosphate

Glycerate 3-phosphate

ADP

ATP

ATP

ADP

Ribulose phosphate

NADP (reduced)

NADP (oxidised)

Glyceraldehyde 3-phosphate
(triose phosphate)

Other carbohydrates, lipids,
proteins and nucleic acids

■ *TIP* Try not to confuse this process with the light-dependent reaction, which requires light and involves the production of ATP, reduced NADP and oxygen.

lignin: a complex *polymer* found in the *cell walls* of certain plant tissues.

■ Lignin is found mainly in *sclerenchyma* and *xylem*. It provides extra mechanical strength to the cell walls, but makes them impermeable to water. Lignification therefore causes the death of the cells, leaving an empty *lumen*. This means that water and dissolved *ions* can easily travel up and down the plant through the lignified tissues.

limiting factor: any environmental factor that limits the rate of a biological process.

■ Blackman's law of limiting factors states that 'when a process is controlled by a number of factors, the factor in least supply will limit the rate of the process'.

■ *e.g.* The graph below shows how light intensity affects the rate of *photosynthesis*.

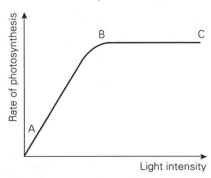

At low light intensity, between the points A and B on the graph, the rate of photosynthesis is limited by the amount of light available. An increase in light intensity will produce an increase in the rate of photosynthesis. Between points B and C, however, increasing the light intensity has no effect on the rate of photosynthesis, so another factor must be limiting. In this case it is probably temperature or the concentration of carbon dioxide which is the limiting factor.

linkage: a term used to describe the relationship between different *genes* located on the same *chromosome*.

Genes on a particular chromosome will normally be inherited together, as a linkage group, unless they are separated during *crossing-over*. Genes located on the *sex chromosomes* will normally be inherited along with sex (*male* or *female*) and are said to be sex-linked.

TIP If during a breeding experiment there is little independent assortment of two or more characteristics, it is likely that the genes controlling these characteristics are linked.

link reaction: the reaction that links *glycolysis* and the *Krebs cycle* during *aerobic respiration*.

In the link reaction, *pyruvate* from glycolysis is converted to *acetylcoenzyme A*, which enters the Krebs cycle. This reaction involves the loss of a molecule of carbon dioxide and the reduction of *NAD*, as summarised by the following equation:

$$\text{pyruvate} + \text{NAD (oxidised)} \longrightarrow \text{acetylcoenzyme A} + CO_2 + \text{NAD (reduced)}$$

lipase: an enzyme that breaks down *lipids* into *fatty acids* and glycerol.

Lipase is a *hydrolase* enzyme produced by the *pancreas* in mammals.

lipid: any of a large group of *organic* substances which are insoluble in water but soluble in organic solvents such as ethanol.

A number of important biological molecules are classified as lipids, including: *triglycerides*, which function as energy stores in plants and animals; *phospholipids*, which are important components of *cell membranes*; *steroids*, which include a number of *hormones*, such as *oestrogen* and *testosterone*. Lipids can combine with proteins to form *lipoproteins*, or with *polysaccharides* to form lipopolysaccharides, as found in the *cell wall* of some *bacteria*.

lipoprotein: a compound consisting of a *lipid* combined with a *protein*.

Lipids are transported in the blood and *lymph* as lipoproteins. For example, *cholesterol* is transported in the form of low-density lipoprotein (LDL). High concentrations of LDL in the blood may predispose an individual to *coronary heart disease*.

liver: a large organ found in vertebrates that performs many important metabolic functions.

The liver carries out many vital processes, many of which have a role in *homeostasis*, as shown by the following list:

- *deamination* of excess *amino acids*
- regulation of the *blood glucose level* by storing *glucose* in the form of *glycogen*

- production of *bile*
- synthesis of *plasma proteins*, such as fibrinogen
- detoxification of poisonous substances
- removal of damaged *red blood cells*
- storage of *iron* and fat-soluble *vitamins*

lock-and-key hypothesis: a model that describes the action of an *enzyme*.

■ The lock-and-key hypothesis proposes that the *active site* of an enzyme has a very specific shape (like a lock) into which the *substrate* molecule (the key) fits exactly, to form an enzyme–substrate complex. This model is generally considered to be less useful than the *induced fit hypothesis*, which argues that the interaction of an enzyme with its substrate causes the active site to change shape slightly in order to allow the formation of an enzyme–substrate complex.

locus: the position on a *chromosome* occupied by a particular *gene*.

■ The two *alleles* of a gene will occupy the same locus on *homologous chromosomes*.

loop of Henle: the hairpin-shaped section of a kidney *nephron*.

■ The loop of Henle is situated between the *proximal convoluted tubule* and the distal convoluted tubule in the nephron. It consists of a thin descending limb, which is permeable to water, and a thick ascending limb, which is impermeable to water. Movement of ions across the walls of the loop of Henle enables it to act as a countercurrent multiplier, building up the concentration of sodium in the *medulla* of the kidney. This process results in the production of concentrated *urine* in the *collecting duct*, as shown in the diagram below.

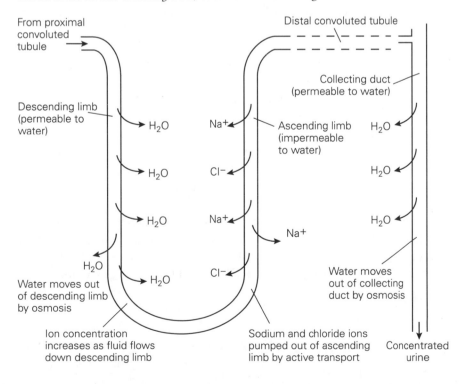

▪ *TIP* A longer loop of Henle will result in the production of more concentrated urine. Therefore animals that live in arid regions, such as the kangaroo rat, have relatively long loops of Henle to conserve water; those living in water-rich regions, such as the beaver, have shorter loops of Henle.

lumen: the space inside a vessel or duct.

▪ *e.g.* The lumen of an *artery* is the space through which blood passes.

lung: one of a pair of gas exchange organs found in air-breathing vertebrates.

▪ The lungs are composed of air ducts (the *trachea, bronchi* and *bronchioles*) and *alveoli*, which provide a very large surface area for gas exchange with the blood.

lung capacity: the total volume of air contained in the *lungs*.

▪ *e.g.* The lung capacity of an average adult male is about five litres or 5000 cm^3.

luteinising hormone (LH): a *hormone* that stimulates *ovulation* in females and the production of *testosterone* in males.

▪ LH is released by the *pituitary gland*. In addition to the functions described above, it also stimulates the development of the *corpus luteum* and the subsequent secretion of *progesterone* in females.

▪ *TIP* Make sure that you know how LH interacts with *oestrogen, progesterone* and *follicle-stimulating hormone (FSH)* to control the *menstrual cycle*.

lymph: the name given to *tissue fluid* after it has drained into the lymphatic system.

▪ Lymph has a similar composition to blood *plasma*, except that it lacks some of the larger proteins. In certain parts of the body, lymph contains large amounts of *fatty acids* and glycerol. These substances are absorbed from the small intestine into lacteals, which are part of the lymphatic system. Lymph eventually rejoins the blood system via the thoracic duct, at a point close to the *heart*.

lymphocyte: a type of *white blood cell* involved in the *immune* response.

▪ There are two types of lymphocyte:
  - *T-cells* are formed in the thymus and are responsible for cellular immunity (the T-cells bind to and destroy foreign *antigens*).
  - *B-cells* are formed in the spleen and lymph nodes and are responsible for humoral immunity (the B-cells secrete *antibodies*, which bind to and destroy foreign antigens).

lysosome: a small membrane-bound *organelle* found in the *cytoplasm* of many cells.

▪ Lysosomes are formed from the *Golgi apparatus* and contain hydrolytic *enzymes*. They are used by certain cells in *phagocytosis*, to break down the ingested material. Lysosomes are also responsible for the autolysis (self-digestion) of cells after death. This is seen in the absorption of the tadpole's tail during the metamorphosis to a frog.

male: an organism that produces male *gametes* (sex cells).

■ Male gametes are usually smaller than *female* gametes and are produced in much larger numbers. They are also mobile, meaning that they can travel to the female gamete in order that *fertilisation* can occur. In mammals, the male gametes are *sperm* and the female gametes are ova (eggs).

maltose: a *disaccharide* found in germinating *seeds*.

■ Maltose is composed of two molecules of α-glucose. It can be synthesised from single *glucose* molecules, or by the *hydrolysis* of *starch* (which consists of long chains of α-glucose).

mass transport: the transport of molecules in bulk from one part of an organism to another.

■ Large organisms need some sort of mass transport system in order to move substances rapidly from one place to another.

■ *e.g.* The blood system in animals and the *phloem* and *xylem* in plants.

■ *TIP* You may read about the mass flow hypothesis in plants. This is a particular case of mass transport and relates to the movement of *organic* substances such as *sucrose* through phloem tissue.

mean: the arithmetic average of a set of numbers.

■ The mean of a set of *n* numbers is obtained by adding all the numbers together and then dividing by *n*.

■ *e.g.* The mean of 25, 27, 31, 32 and 35 is (25 + 27 + 31 + 32 + 35)/5 = 30.

mechanoreceptor: a sensory *receptor* that responds to a mechanical *stimulus*, such as touch, sound or pressure.

■ *e.g.* Mechanoreceptors in the skin detect touch, while others in the ear are concerned with balance and hearing.

medulla: the inner part of the *kidney*.

■ The medulla contains the *loops of Henle* and *collecting ducts* of the kidney *nephrons*, together with a large number of blood vessels.

medulla oblongata: part of the *brain* which is responsible for regulating the breathing rate and blood flow around the body.

■ The medulla regulates blood flow by controlling heart rate, blood pressure and local *vasodilation* and *vasoconstriction* of *arterioles*.

meiosis: a type of cell division used in the production of *gametes*.

■ Meiosis involves two successive divisions of a *diploid* cell, resulting in the production of *haploid* gametes. This is summarised in the table below.

Stage	1st division	2nd division
Prophase	*Homologous chromosomes* form pairs and may exchange genetic material during *crossing-over*.	Individual chromosomes can be seen to consist of a pair of *chromatids*.
Metaphase	Homologous pairs line up at random on the equator of the cell, attached to spindle fibres by their *centromeres*.	Individual chromosomes line up at random on the equator of the cell, attached to spindle fibres by their centromeres.
Anaphase	Homologous chromosomes separate and move to opposite ends of the dividing cell.	Chromatids separate and move to opposite ends of the dividing cell.
Telophase	The first division is completed.	The second division is completed and four gametes are produced from the original parent cell.

■ *TIP* Try not to confuse meiosis (reduction division) with *mitosis*, which is a type of cell division used for *growth*. The words meiosis and mitosis even look similar, so make sure that your spelling is correct. Finally, it is useful to note that the processes involved in the second meiotic division are essentially the same as those seen in mitosis, but with only half the normal number of chromosomes.

menstrual cycle: the cycle of events associated with ovulation and the development and breakdown of the *endometrium*.

■ In humans, the menstrual cycle averages 28 days in length. It is regulated by four *hormones*, as summarised in the diagram below.

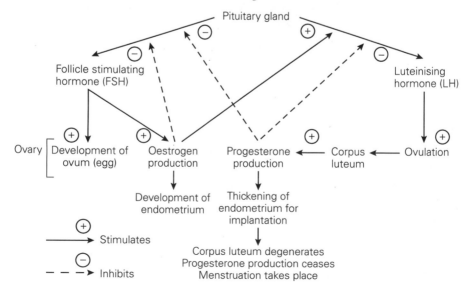

If *implantation* occurs, the secretion of *progesterone* is taken over by the *placenta* and menstruation does not take place.

meristem: a type of plant *tissue* in which *mitosis* takes place.

■ The meristem gives rise to cells that form new tissues by *differentiation*. Apical meristems are found at the tips of shoots and roots and are responsible for vertical growth. Lateral meristems are involved in secondary thickening and are responsible for horizontal growth.

messenger RNA (mRNA): a type of *RNA* which carries genetic information from the *nucleus* to the *cytoplasm* during *protein synthesis*.

■ mRNA is a long, single-stranded molecule which carries the *genetic code* in the form of *codons*. The code is copied from *DNA* to mRNA during *transcription* in the nucleus. mRNA then moves to the *ribosomes*, where proteins are synthesised by *translation*. This process is summarised in the diagram below.

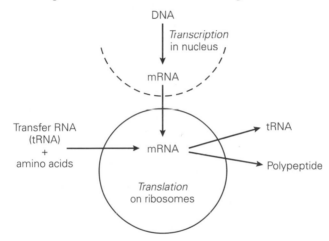

■ *TIP* Remember that the sequence of bases in mRNA is complementary to those in DNA, except that uracil (U) replaces thymine (T). For example, the DNA sequence AACTTAGGT would be transcribed into UUGAAUCCA.

metabolism: the sum total of all the biochemical reactions taking place within an organism.

■ Metabolic reactions are controlled by *enzymes* and can be divided into two groups. Anabolism involves synthetic reactions which require energy, such as *protein synthesis*. Catabolism involves breakdown reactions which release energy, such as *respiration*. Substances involved in metabolism are called metabolites and the sequence of biochemical reactions is known as a metabolic pathway.

micronutrient: nutrients required in very small quantities in order to maintain health.

■ *e.g.* *Vitamins*, such as vitamin A and vitamin C, and mineral *ions*, such as *iron* and *calcium*.

microorganism: any organism that can be observed only with the aid of a microscope.

■ *e.g.* Bacteria, viruses and some *protoctists*.

mitochondrion: a rod-shaped *organelle* found in the *cytoplasm* of *eukaryotic* cells.

■ The function of mitochondria (plural) is to carry out the chemical reactions involved in the *Krebs cycle* and the *electron transport chain* of *aerobic respiration*. The diagram below shows the general structure of a mitochondrion.

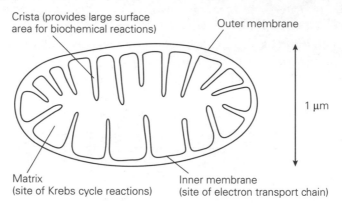

Crista (provides large surface area for biochemical reactions)

Outer membrane

1 μm

Matrix (site of Krebs cycle reactions)

Inner membrane (site of electron transport chain)

■ *TIP* Cells with large energy demands, such as *muscle* cells or those carrying out *active transport*, tend to contain large numbers of mitochondria.

mitosis: a type of cell division used for *growth*.

■ Mitosis involves the production of genetically identical daughter cells, which have the same *chromosome* number as the parent cell. The process of mitosis is summarised in the table below.

Stage	Key events
Prophase	Chromosomes shorten and thicken. The nuclear membrane starts to break down and the spindle forms.
Metaphase	Chromosomes line up on the equator of the cell attached to spindle fibres by their *centromeres*.
Anaphase	*Chromatids* separate and move to opposite ends of the dividing cell. Each chromatid has now become a chromosome.
Telophase	The spindle fibres break down, the nuclear membrane re-forms, chromosomes elongate and cell division is completed.

■ *TIP* Try not to confuse mitosis with *meiosis* (reduction division), which is a type of cell division used for the production of *gametes*. The words meiosis and mitosis even look similar, so make sure that your spelling is correct.

molecule: a substance formed by the combination of *atoms*.

■ The atoms comprising a molecule may be the same, as in nitrogen gas (N_2), or different, as in carbon dioxide (CO_2). Some biological molecules, such as *DNA*, contain hundreds or thousands of atoms and are known as macromolecules.

monocotyledon: a flowering plant that produces seeds with only one *cotyledon* (seed leaf).

◾ Monocotyledons share the following features:
- the leaves have parallel veins
- the vascular bundles are scattered throughout the stem
- the flowers are in three parts, or multiples of three

◾ *e.g.* Orchids and wheat are monocotyledons.

◾ *TIP* Make sure that you know the differences between monocotyledons and *dicotyledons*, which produce seeds with two cotyledons.

monocyte: a type of *white blood cell*.

◾ Monocytes ingest *bacteria* and cell debris in the blood by *phagocytosis*.

monohybrid inheritance: the inheritance of a single characteristic.

◾ The different forms of the characteristic are controlled by different *alleles* of the same *gene*.

◾ *e.g.* A monohybrid cross between two pure-breeding pea plants, one with yellow seeds and one with green seeds, would be expected to produce an F_1 *generation* with only yellow seeds. This is because yellow coloration is dominant, and green coloration is recessive. If these F_1 plants are interbred, the F_2 generation will have a mixture of yellow and green seeds in a ratio of 3:1. This monohybrid cross is summarised in the diagram below.

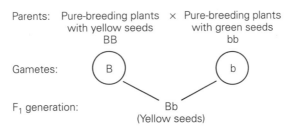

Let the alleles be represented by the following letters:
B = yellow; b = green

Parents: Pure-breeding plants × Pure-breeding plants
with yellow seeds with green seeds
BB bb

Gametes: B b

F_1 generation: Bb
(Yellow seeds)

Interbreeding the F_1 generation:

		Male gametes	
		B	b
Female gametes	B	BB yellow	Bb yellow
	b	Bb yellow	bb green

3 yellow seeds : 1 green seed

■ *TIP* It is a good idea to practise writing out genetic crosses, as in the example above. Try to remember the following points:

- set your work out clearly, labelling the parents, gametes and new generations
- do not confuse the terms *genotype* and *phenotype*, or *gene* and *allele*
- do not represent alleles with letters that look similar in capitals and lower case, such as S and s, or K and k
- remember that gametes are *haploid*, so they only have one of each pair of alleles

monomer: a *molecule* that can combine with similar molecules to form a *polymer*.

■ A monomer represents the single sub-unit of a polymer.

■ *e.g.* *Polysaccharides* are made up of *monosaccharides*; proteins are built up from *amino acids*; and *DNA* is composed of *nucleotides*.

monosaccharide: the simplest form of *carbohydrate*.

■ Monosaccharides are known as single sugars and have the general formula $(CH_2O)_n$. If $n = 3$ the monosaccharide is a triose, e.g. triose phosphate (an important intermediate in the *light-independent reaction* of *photosynthesis*). If $n = 5$ the monosaccharide is a *pentose*, e.g. ribose (a component of *RNA*). If $n = 6$ the monosaccharide is a *hexose*, e.g. *glucose* (the major respiratory substrate of most organisms).

■ *TIP* Monosaccharides can be built up into *disaccharides* and *polysaccharides* by *condensation* reactions.

mRNA: see *messenger RNA*.

mucus: a *glycoprotein* which has lubricating and protective functions.

■ Mucus acts as a lubricant in *synovial joints* and in the digestive tract. It also protects the lining of the gut from damage caused by digestive juices and coats the surface of the *trachea*, *bronchi* and *bronchioles*, to trap dust and *micro-organisms*.

multiple alleles: a condition in which a single *gene* has three or more *alleles* controlling a particular characteristic.

■ *e.g.* ABO *blood groups* in humans are controlled by a single gene with three alleles, I^A, I^B and I^o. Each person can only possess two of these alleles. I^A and I^B are *codominant* and are both *dominant* to I^o. Therefore, the possible *genotypes* and *phenotypes* (blood groups) are as follows:

- $I^A I^A$ or $I^A I^o$ = blood group A
- $I^B I^B$ or $I^B I^o$ = blood group B
- $I^A I^B$ = blood group AB
- $I^o I^o$ = blood broup O

muscle: a *tissue* consisting of contractile fibres that can produce movement.

■ There are three types of muscle — skeletal, smooth and cardiac. *Skeletal muscle* is attached to the *skeleton* and produces voluntary movement at joints. *Smooth muscle* is responsible for involuntary movement, e.g. *peristalsis* in the gut. *Cardiac muscle* exists only in the *heart*.

■ *TIP* Muscle is an example of an *effector*.

mutation: any change in the amount or arrangement of *DNA* in a cell.

■ There are two types of mutation — point (gene) and chromosome. *Point (gene) mutations* cause changes in the sequence of bases in DNA due to the deletion, insertion or substitution of a base. An example of a condition caused by a point mutation is sickle-cell anaemia. *Chromosome mutations* cause changes in the structure or number of chromosomes. Changes in structure may be due to the deletion or translocation of part of a chromosome. Changes in number may be due to *polysomy* or *polyploidy*. An example of a condition caused by a chromosome mutation is *Down's syndrome*. Mutations occur naturally at a low rate, but this rate may be increased by mutagens, such as *ionising radiation*. Most mutations are harmful, but a few may increase the fitness of an organism and so spread through the *population* by *natural selection*. Mutation is therefore the source of new genetic variation and is essential for *evolution*.

mutualism: a relationship between two different *species* which results in benefit to both.

■ *e.g.* Hermit crabs and sea anemones often have a mutualistic relationship. The crab gains camouflage and protection from the anemone which lives on its shell and the anemone gains scraps of food dropped by the crab.

mycelium: a network of *hyphae* that forms the body of many *Fungi*.

myelin: a mixture of *phospholipid* and *cholesterol* which forms a sheath around many nerve cells.

■ Myelin is produced by *Schwann cells* and is laid down as insulating blocks along the length of the *neurone*. Points at which there is no myelin sheath are known as nodes of Ranvier. Myelin speeds up the conduction of an *impulse*, as it allows the impulse to jump from one node to the next. This is known as *saltatory conduction*.

myofibrils: the microscopic fibres that make up the larger fibres of *skeletal muscle*.

■ Myofibrils give skeletal muscle fibres their characteristic striated or striped appearance, due to the presence of alternating light and dark bands. These bands are due to the regular arrangement of filaments made from the proteins *actin* and *myosin*, as shown in the diagram below.

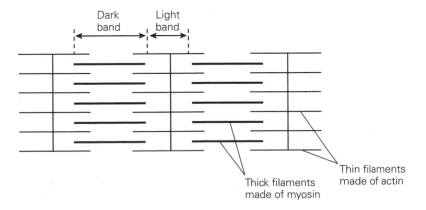

When the myofibril contracts, the actin and myosin filaments slide past one another, according to the *sliding-filament hypothesis*.

myoglobin: an iron-containing protein found in *muscle*.

■ Myoglobin consists of a *polypeptide* chain linked to a *haem* group. Its function is to store oxygen in muscle tissue for use during strenous exercise. Myoglobin has to obtain its oxygen from *haemoglobin* in the blood. It therefore has a high affinity for oxygen and its *oxygen dissociation curve* lies to the left of that of haemoglobin.

myosin: a protein that forms the thick filaments in the microscopic fibres of *skeletal muscle*.

■ *TIP* This term is often seen in conjunction with *actin*, a protein forming the thin filaments found in the *myofibrils* (microscopic fibres) of skeletal muscle.

NAD: a molecule that acts as a hydrogen acceptor during *respiration*.

■ NAD (nicotinamide adenine dinucleotide) is a *coenzyme* which forms part of the *electron transport chain* in *aerobic respiration*. It acts as a hydrogen acceptor during *glycolysis*, the *link reaction* and the *Krebs cycle*, becoming reduced NAD. It is then re-oxidised by passing a hydrogen atom (or, more accurately, a proton and an electron) to the next carrier in the electron transport chain. This reaction releases energy which is used to produce *ATP*.

■ *TIP* Try not to confuse NAD with *NADP*, which performs a similar role during *photosynthesis*.

NADP: a molecule that acts as a hydrogen acceptor during *photosynthesis*.

■ NADP (nicotinamide adenine dinucleotide phosphate) is a *coenzyme* which acts as a hydrogen acceptor during the *light-dependent reaction*, becoming reduced NADP. It is then re-oxidised by passing a hydrogen atom to glycerate 3-phosphate in the *light-independent reaction*. This reaction produces glyceraldehyde 3-phosphate which can be built up into *carbohydrates* and other *organic* molecules.

■ *TIP* Try not to confuse NADP with *NAD*, which performs a similar role during *respiration*.

natural selection: the mechanism responsible for the *evolution* of new *species*.

■ The process of natural selection can be summarised as follows.

- All species tend to produce many more offspring than can ever survive. However, the size of populations tends to remain more or less constant, meaning that most of these offspring must die. It also follows that there must be competition for resources such as mates, food and territories. Therefore, there is a 'struggle for existence' among individuals.

- Individuals within a species differ from one another (*variation*) and many of these differences are inherited (genetic variation).

- Competition for resources, together with variation between individuals, means that certain members of the population are more likely to survive and reproduce than others. These individuals will inherit the characteristics of their parents and evolutionary change will take place through natural selection.

n

■ *TIP* Natural selection exists in three forms – *directional selection, disruptive selection* and *stabilising selection*. It is useful to know the differences between these forms of selection.

negative feedback: a mechanism used in *homeostasis* for maintaining optimum conditions in the body.

■ If a physiological factor deviates from its optimum, the deviation is sensed and negative feedback mechanisms are used to return the factor to its optimum. This idea is summarised in the diagram below.

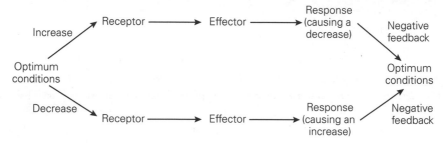

■ *e.g.* If the *blood glucose level* in the body rises, *insulin* is secreted from the *pancreas* which causes the glucose level to fall back to the optimum level.

■ *TIP* Note that in cases of *positive feedback*, a deviation from the optimum causes a greater deviation to occur.

nephron: the functional unit of the *kidney*.

■ Each kidney contains over one million nephrons, which regulate the composition of blood and the formation of *urine*. The general structure of a nephron is shown below.

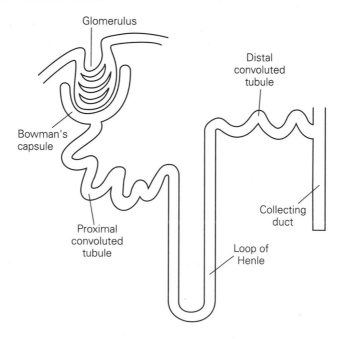

Fluid enters *Bowman's capsule* by *ultrafiltration* from the blood. This *glomerular filtrate* consists of blood *plasma* minus the large *plasma proteins*. Useful substances, such as *glucose, amino acids* and some salts and water, are reabsorbed back into the blood as the filtrate passes through the nephron. The end product is a relatively concentrated solution of *urea*, salts and water, known as urine, which passes out of the body via the bladder.

net productivity: a measure of the rate at which energy is used to form new tissues.

■ *Organic* molecules accumulate in living organisms by *autotrophic* or *heterotrophic* nutrition. The total amount of material accumulated is known as the gross productivity. However, some of this material is used for *respiration*, and some is lost by *excretion* and defaecation. Net productivity is the amount of organic material remaining for *growth*, as summarised by the following equation:

net productivity = gross productivity − (respiration + excretion + defaecation)

Net productivity is normally measured for a particular area over a particular time and in units of energy, e.g. $kJ\,m^{-2}\,year^{-1}$.

neuromuscular junction: the point at which a motor *neurone* connects with a *muscle*.

■ Impulses are transmitted from the motor neurone to the muscle by *acetylcholine* (a *neurotransmitter*). Acetylcholine is released from the end of the neurone and depolarises the surface membrane of a muscle fibre. This triggers an *action potential* in the muscle, leading to contraction.

■ *TIP* The action of a neuromuscular junction is very similar to that of a *synapse*.

neurone: a cell specialised for the conduction of nerve *impulses*.

■ Neurones are well-adapted for conducting impulses, having an elongated shape and specialised structures to form connections with *receptors, effectors* and other neurones via *synapses*. Many neurones are coated with *myelin*, which helps to speed up the conduction of nerve impulses. There are three basic types of neurone. Sensory neurones transmit impulses from a receptor to the *central nervous system*. Motor neurones transmit impulses from the central nervous system to an effector. Intermediate neurones connect sensory and motor neurones within the central nervous system.

neurotransmitter: a chemical that permits the transmission of a nerve *impulse* across a *synapse*.

■ The neurotransmitter is released from a synaptic knob at the tip of an *axon* into the synaptic cleft. It diffuses across to the postsynaptic membrane, where it initiates an *action potential* in the next neurone. Neurotransmitters also act at *neuromuscular junctions*, where they stimulate muscle contraction.

■ *e.g.* Noradrenaline and *acetylcholine*.

niche: the role of a *species* in an *ecosystem*.

■ Each species has its own niche, which includes its location, feeding relationships and behaviour.

nicotinamide adenine dinucleotide: see *NAD*.

n

nicotinamide adenine dinucleotide phosphate: see *NADP*.

nitrification: the process by which *bacteria* convert ammonium compounds into nitrites and nitrates in the soil.

■ Nitrifying bacteria, such as *Nitrosomonas* and *Nitrobacter*, oxidise ammonium compounds and nitrites respectively, releasing energy. The bacteria make use of this energy to produce *organic* compounds from simple *inorganic* molecules by *chemosynthesis*. These chemical reactions are important in the *nitrogen cycle*.

■ *TIP* The reactions involved in nitrification are essentially the reverse of those that comprise *denitrification*, as summarised by the nitrogen cycle. Do not confuse nitrification with *nitrogen fixation*, which is the conversion of atmospheric nitrogen into nitrogen-containing compounds.

nitrogen cycle: one of several *nutrient cycles*, describing how the element nitrogen cycles in the environment.

■ The nitrogen cycle is summarised in the diagram below.

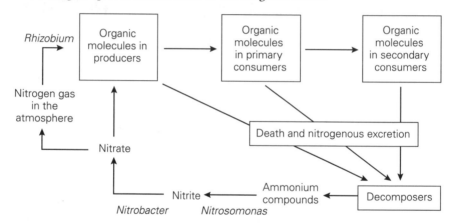

■ *TIP* Note that *denitrification* and *nitrogen fixation* are not always involved in the nitrogen cycle.

nitrogen fixation: the process by which atmospheric nitrogen is converted into nitrogen-containing compounds.

■ *Bacteria* are the only organisms capable of fixing nitrogen. They live either freely in the soil, such as *Clostridium*, or in a *mutualistic* relationship with *leguminous plants*, such as *Rhizobium*. The ammonia produced by nitrogen fixation can be converted to nitrate (see *nitrogen cycle*) and used by plants. Some atmospheric nitrogen is fixed by the action of electrical discharge (lightning), the nitrogen oxides formed being washed into the soil by rain and then taken up by plants as nitrates. Finally, nitrogen can be added to the soil artificially by applying *fertilisers*.

■ *TIP* Do not confuse nitrogen fixation with *nitrification*, which is the conversion of ammonium compounds into nitrites and nitrates.

non-competitive inhibition: the inhibition of *enzyme* activity due to the binding of an inhibitor molecule at a site other than the *active site*.

■ The binding of a non-competitive inhibitor changes the shape of the active site, preventing the enzyme from functioning and reducing the rate of reaction. Unlike *competitive inhibition*, non-competitive inhibition cannot be reversed by increasing the concentration of *substrate*.

■ *TIP* Try not to confuse this term with competitive inhibition, which is a reduction in the rate of an enzyme-catalysed reaction by an inhibitor molecule that is similar in shape to the normal substrate.

non-disjunction: the failure of a pair of *homologous chromosomes* to separate during *meiosis*.

■ Non-disjunction usually results in the production of a *gamete* with one more or one less chromosome than normal. If this gamete is fertilised by a normal gamete, the resulting *zygote* will suffer from *polysomy* (a type of *chromosome mutation*). *Down's syndrome* is an example of a condition caused by non-disjunction.

noradrenaline: a chemical that allows the transmission of an *impulse* from one *neurone* (nerve cell) to another.

■ Neurones connect with each other at a *synapse*. The arrival of a nerve impulse at the presynaptic membrane causes the release of noradrenaline, which diffuses across the synaptic cleft to receptors on the postsynaptic membrane. This triggers another nerve impulse in the postsynaptic neurone.

nucleic acid: a biological molecule consisting of a long chain of *nucleotides*.

■ *e.g.* Ribonucleic acid (RNA) and *deoxyribonucleic acid (DNA)*.

nucleolus: part of the *nucleus* of a cell where ribosomal *RNA* is produced.

■ The nucleolus plays an important role in the synthesis of *ribosomes* and is therefore vital for *protein synthesis*.

nucleotide: the basic unit from which *nucleic acids* are formed.

■ A nucleotide consists of a *pentose* (5-carbon) sugar, a nitrogenous base and a phosphate group, as shown in the diagram below.

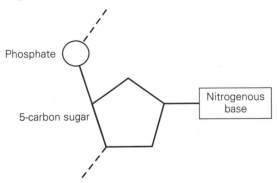

The pentose sugar may be either *ribose* or deoxyribose, and the nitrogenous base may be *adenine, cytosine, guanine, thymine* or *uracil*. Nucleotides may join together by *condensation* reactions to form dinucleotides, such as *NAD*, or *polynucleotides*, such as *DNA* or *RNA*.

n

nucleus: a large *organelle* found in *eukaryotic cells*.

■ The nucleus contains *DNA* in the form of *chromosomes* and is the site of synthesis of *RNA*. It is separated from the rest of the cell by a double membrane (envelope), which has pores to allow the movement of substances in or out. For example, *messenger RNA (mRNA)* passes out of the nucleus during *protein synthesis*.

null hypothesis: a hypothesis which is based on the assumption that there is no significant difference between two sets of data.

■ It is unlikely that there will be no difference between two sets of data, but a null hypothesis predicts that there will be no significant difference, or that any difference will be due to chance. A statistical test can be used to test a null hypothesis. If the test indicates a significant difference between two sets of data, the null hypothesis is rejected. Otherwise the null hypothesis is accepted.

nutrient cycling: the way in which *elements* move through *ecosystems*.

■ The basic features of a nutrient cycle are the same for all elements. The element is taken up in an *inorganic* form and incorporated into *organic* molecules by *producers*. It is then passed along a *food chain* to *primary consumers* and secondary consumers. Death of any of these organisms results in their breakdown by *decomposers*, which releases the element in an inorganic form and completes the cycle.

■ *e.g.* The *carbon cycle* and the *nitrogen cycle*.

oesophagus: the section of the digestive system that lies between the mouth and the *stomach*.
- The oesophagus is a muscular tube whose function is to transfer food to the stomach by *peristalsis*.

oestrogen: a sex *hormone* produced by the *ovaries* in females.
- Oestrogen plays an important role in the regulation of the *menstrual cycle*, the maintenance of *pregnancy* and the development of female secondary sexual characteristics, such as breasts.
- **TIP** Make sure that you know how oestrogen interacts with *follicle-stimulating hormone (FSH)*, *progesterone* and *luteinising hormone (LH)* to control the menstrual cycle.

oogenesis: the production of female *gametes* (eggs) by the *ovary*.
- Certain cells (oogonia) within the ovary divide by *mitosis* to produce large numbers of potential gametes called primary oocytes. The primary oocytes then undergo *meiosis* to produce first a secondary oocyte and then an *ovum* (egg). Oogenesis is stimulated by *follicle-stimulating hormone (FSH)*.

organ: part of an organism that is specialised to perform a particular function(s).
- Organs are made up of a number of different *tissues*. Several organs may work together in an organ system, such as the digestive system.
- **e.g.** Eyes, *liver* and *kidneys* in animals; leaves, *roots* and *flowers* in plants.
- **TIP** It is useful to know the different levels of organisation in living organisms, starting at the simplest (smallest) level: *organelle — cell — tissue —* organ — organ system — organism.

organelle: a structure found in the *cytoplasm* of a cell which has a specific function.
- Most organelles are separated from the cytoplasm by one or two *cell membranes*.
- **e.g.** Lysosomes, *chloroplasts* and *ribosomes*.

organic: any *molecule* based on carbon.
- Some organic molecules found in living organisms may be very large, such as *proteins*, *lipids* and *carbohydrates*.
- **TIP** All molecules can be classified as either organic or *inorganic* (not based on carbon).

ornithine cycle: a series of biochemical reactions that convert ammonia into *urea*.

■ The ornithine cycle takes place in the *liver* of mammals. *Deamination* of excess *amino acids* produces the highly toxic nitrogenous waste product, ammonia. This is converted to less toxic urea by the reactions of the ornithine cycle, as shown in the diagram below.

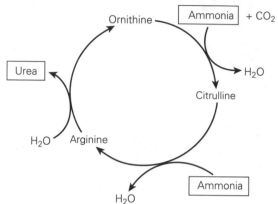

■ *TIP* Note that urea is produced in the *liver*, but is excreted from the body by the *kidneys*.

osmoregulation: the ability of an organism to regulate the *water potential* of its body fluids.

■ Osmoregulation is an example of *homeostasis*. The process involves monitoring the water potential inside the body and responding to changes by excreting or absorbing and retaining water (and salts). Freshwater organisms tend to gain water by *osmosis* and must therefore eliminate the excess. This is done by various mechanisms, such as the use of *contractile vacuoles* by *protoctists* and *kidneys* in freshwater fish. Marine organisms have the opposite problem (although many of them are *isotonic* to their environment), using various mechanisms to prevent excessive water loss and enhance the *excretion* of salts. Terrestrial organisms risk desiccation and therefore have particularly long *nephrons* to increase water retention.

osmosis: the *diffusion* of water molecules across a *partially permeable membrane*.

■ During osmosis, water molecules diffuse from an area of high *water potential* to one of lower water potential (down a water potential gradient). Pure water has the highest water potential (zero) and, as the solute concentration increases, the water potential of a solution decreases (becomes more negative). The process of osmosis is summarised in the diagram (right).

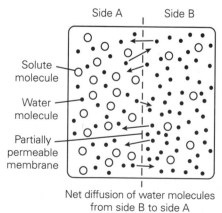

Net diffusion of water molecules from side B to side A

■ *TIP* Remember that *cell membranes* are partially permeable and that osmosis is important for moving water through the bodies of living organisms. Do not forget that only water moves by osmosis, nothing else.

ovary: an *organ* that produces female *gametes*.

■ In plants, each ovary contains a number of ovules, which become *seeds* following *fertilisation*. The ovary itself develops into a *fruit*. In vertebrates, the female gametes are known as ova (eggs) and the ovaries are also responsible for the production of *oestrogen* and *progesterone*.

ovum: a *gamete* produced by female vertebrates.

■ An ovum (egg cell) is produced by an *ovary*.

oxidation: the removal of an *electron* from a molecule.

■ Oxidation is also used to describe the removal of hydrogen or the addition of oxygen to a molecule. The oxidation of one molecule is always accompanied by the *reduction* of another. In biological systems, oxidation and reduction are controlled by *enzymes* called *oxidoreductases*.

■ *TIP* Make sure that you do not confuse oxidation and reduction, which is the gain of an electron by a molecule. Remember OILRIG (Oxidation Is Loss Reduction Is Gain).

oxidative phosphorylation: the process of *ATP* production by the *electron transport chain*.

■ Oxidative phosphorylation involves *oxidation* coupled to the phosphorylation of ADP (to make ATP).

■ *TIP* Try not to confuse this process with *photophosphorylation*, which is the production of ATP using light energy and takes place during the *light-dependent reaction* of *photosynthesis*.

oxidoreductase: a class of *enzymes* that catalyse biochemical reactions involving *oxidation* and *reduction*.

■ Oxidoreductases catalyse the transfer of hydrogen, oxygen or electrons between molecules. Several oxidoreductases are involved in the metabolic pathways of *respiration*.

■ *e.g.* Cytochrome oxidase is an enzyme which transfers electrons from *cytochromes* to oxygen (forming water) at the end of the *electron transport chain*.

oxygen debt: a deficit of oxygen that is caused by *anaerobic respiration* in the body.

■ Anaerobic respiration in muscle tissue produces lactic acid, a toxic product that must eventually be oxidised back to *pyruvate*. The oxygen required for this process represents the oxygen debt. After a period of strenuous activity (when the muscles have been respiring anaerobically), we continue to breathe heavily in order to supply oxygen to 'pay off' the oxygen debt.

oxygen dissociation curve: a curve describing the relationship between oxygen concentration and the percentage saturation with oxygen of a respiratory pigment such as *haemoglobin*.

O

■ Oxygen concentration is usually represented as the partial pressure of oxygen (pO_2). The diagram below shows a typical oxygen dissociation curve for human haemoglobin.

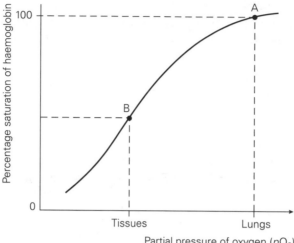

The steep rise of the curve at low partial pressures of oxygen indicates that a small change in pO_2 results in a relatively large change in the saturation of haemoglobin with oxygen. Therefore in the lungs, where the pO_2 is high (A), the blood is rapidly saturated with oxygen. Similarly, in rapidly respiring tissues where the pO_2 is low (B), oxygen is released from haemoglobin for use by the tissues. The affinity of haemoglobin for oxygen depends upon the concentration of carbon dioxide in the blood, as described by the *Bohr effect*.

■ *TIP* If the curve moves to the left, haemoglobin will have a higher affinity for oxygen. Likewise, if it moves to the right, haemoglobin has a lower affinity for oxygen. Try to remember that a shift to the **R**ight makes it more likely that haemoglobin will **R**elease oxygen.

palisade cell: a type of cell found just beneath the upper *epidermis* of a leaf.

■ The structure of a palisade cell is shown in the diagram below.

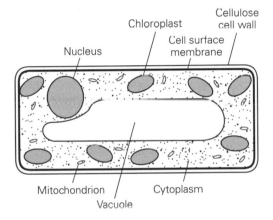

Palisade cells are elongated at right angles to the leaf surface and are packed closely together for maximum absorption of light. They contain large numbers of *chloroplasts* and their principal function is *photosynthesis*.

pancreas: a secretory organ found just below the *stomach* in vertebrates.

■ The pancreas has two main secretory functions.

- An *exocrine* function: pancreatic juice is secreted into the *duodenum* via the pancreatic duct. It is an alkaline solution of digestive *enzymes,* such as *amylase, lipase* and *trypsin*.

- An *endocrine* function: the *islets of Langerhans* secrete *glucagon* and *insulin* into the blood.

parasitism: a relationship between two different *species* which results in benefit to one (the parasite) and harm to the other (the host).

■ *e.g.* Leeches (parasites) temporarily attach themselves to the outer surface of a number of hosts and suck blood.

parasympathetic nervous system: the part of the *autonomic nervous system* which regulates physiological functions when the body is at rest.

■ The *neurotransmitter* used in the parasympathetic nervous system is *acetylcholine*.

Examples of parasympathetic stimulation include a decrease in ventilation rate, constriction of pupils and a decrease in cardiac output.

■ *TIP* Try not to confuse the parasympathetic system with the *sympathetic nervous system*, which is the part of the autonomic nervous system which regulates physiological functions when the body is active. The two systems generally have antagonistic effects.

parenchyma: a type of plant *tissue*.

■ Parenchyma is composed of relatively unspecialised living cells with thin *cellulose cell walls*. They are generally *turgid* to support the aerial parts of the plant and can be used for storage of materials.

partially permeable membrane: a membrane that is permeable to small molecules but does not allow the passage of larger molecules.

■ All *cell membranes* are partially permeable, allowing water and small solute molecules to pass through, but not large solute molecules. This is important in *osmosis* (the *diffusion* of water molecules across a partially permeable membrane).

passive transport: movement of molecules across *cell membranes* that does not require energy.

■ *e.g.* Diffusion, *facilitated diffusion* and *osmosis*.

PCR: see *polymerase chain reaction*.

pectin: a *polysaccharide* found in plant *cell walls*.

■ Pectin is normally found in an insoluble form in cell walls and in the middle lamella between cell walls, where it helps to cement *cellulose* fibres together. However, in ripening fruits it changes into a soluble form, as indicated by the softening of the tissues. *Enzymes* which digest pectin are used commercially in the extraction of fruit juices.

pentose: a *monosaccharide* containing five carbon atoms.

■ *e.g.* Ribose, deoxyribose and ribulose are all *pentoses*.

pepsin: a protein-digesting *enzyme* produced by the *stomach*.

■ Pepsin is secreted in *gastric juice* in an inactive form, pepsinogen, to avoid self-digestion by the secretory cells. It is activated to pepsin by the hydrochloric acid present in gastric juice. Pepsin is an *endopeptidase*, hydrolysing the *peptide bonds* within proteins and converting them into shorter *polypeptides*.

peptide bond: the type of bond that links *amino acids* together in a *polypeptide*.

■ A peptide bond is formed by a *condensation reaction* between two amino acids, as shown in the diagram below.

Peptide bond

$$H-N-C-C+OH- + H+N-C-C-OH \longrightarrow H-N-C+C-N+C-C-OH- + H_2O$$

amino acid 1   amino acid 2   dipeptide   water

peristalsis: the movement of material along a tube by waves of muscular contraction.

■ During peristalsis, *smooth muscle* contracts and relaxes in a wave-like manner, squeezing material along the tube.

■ *e.g.* Peristalsis is responsible for moving material through the small and large intestines.

pesticide: any chemical compound that is used to kill pests.

■ A pest is defined as an organism that destroys agricultural production or is harmful to humans. Pesticides include *herbicides*, which kill unwanted plants or weeds; insecticides, which kill insect pests; and fungicides, which kill fungi. The disadvantages of pesticides are:

- they may be toxic to non-target organisms
- they may be non-biodegradable, so that they persist in the environment
- they may *bioaccumulate* in living organisms
- the frequent use of a pesticide may result in the pest population becoming resistant to its effects

■ *TIP* Pesticides are not the same as *fertilisers*, which are substances added to soil in order to increase its productivity.

petal: one of the outer parts of a *flower* that makes up the *corolla*.

■ Petals of insect-pollinated plants are usually brightly coloured and often scented. Petals of wind-pollinated plants are usually very small or absent.

pH: a measure of the acidity or alkalinity of a solution.

■ pH is inversely related to the concentration of hydrogen *ions* in a solution. It is measured on a scale of 0 to 14, where 0 indicates a very strong acid, 14 indicates a very strong alkali and 7 is neutral. In other words, a low pH (0–2) indicates a high concentration of hydrogen ions (a strong acid), and a high pH (12–14) indicates a low concentration of hydrogen ions (a strong alkali).

phagocytosis: the process by which cells take up large particles from their environment.

■ Phagocytosis is an example of *endocytosis*. The cell surface membrane invaginates to capture the particles and then encloses around them to form a *vacuole*. The contents of the vacuole are then broken down by the action of *enzymes* from *lysosomes*. Phagocytosis is the mechanism used by certain *protoctists* to acquire nutrients and it forms an important defence mechanism in animals, because certain *white blood cells* are capable of engulfing foreign material such as *bacteria*.

■ *TIP* Try not to confuse this process with *pinocytosis*, which is the process by which cells actively take up liquids from their environment.

phenotype: the physical expression of the *genotype* of an organism.

■ The genotype of an organism is its genetic composition. For example, a pea plant may contain *alleles* which influence its height. In this case, the phenotype (appearance) of the pea plant could be either tall or short. It is important to note that the *environment* may have significant effects on the phenotype of an organism. For example, a pea plant may have the required genetic composition to be tall, but without appropriate amounts of light and water it will not reach its maximum height.

phloem: the plant *vascular* tissue concerned with the transport of soluble *organic* molecules, such as *sucrose* and *amino acids*.

■ In general, phloem transports organic molecules from the site of production in the leaves (the source) to the other parts of the plant (the sink). It is thought that the molecules move by a system of mass flow, although the precise mechanism is unknown.

phospholipid: a type of *lipid* which contains a phosphate group.

■ The diagram below shows the general structure of a phospholipid molecule.

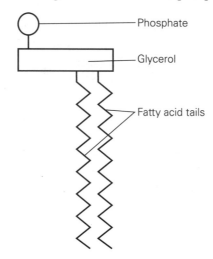

The phosphate 'head' of the molecule is hydrophilic (water soluble) and the fatty acid 'tail' is hydrophobic (insoluble in water). This property of phospholipids means that they form a bilayer in aqueous solutions, an important factor in the formation of *cell membranes*.

photoautotroph: an organism that synthesises complex *organic* molecules from simple *inorganic* molecules, using light energy.

■ All organisms that are capable of *photosynthesis* are photoautotrophs. They act as *producers* in *food chains* and *food webs*.

photolysis: the splitting of water molecules that takes place during *photosynthesis*.

■ Photolysis occurs during the *light-dependent reaction* and can be summarised by the following reaction:

$$2H_2O \longrightarrow 4H^+ + 4e^- + O_2$$

water hydrogen ions electrons oxygen

The electrons are used to replace those lost by *chlorophyll* during non-cyclic *photophosphorylation*. The hydrogen ions are used to reduce *NADP*, for use in the *light-independent reaction*. The oxygen is released as a waste product of photosynthesis.

p

photophosphorylation: the synthesis of *ATP* using light energy during *photosynthesis*.

■ Photophosphorylation occurs during the *light-dependent reaction* of photosynthesis. Light striking a *chlorophyll* molecule causes a high energy electron to be released and passed along an *electron transport chain*. As the electron is transferred from one carrier to the next, energy is lost and some of this is used to make ATP.

■ *TIP* Try not to confuse this process with *oxidative phosphorylation*, which is the production of ATP using the chemical energy from *organic* molecules and takes place during *aerobic respiration*.

photoreceptor: a cell which is sensitive to light.

■ Most photoreceptors contain a *pigment* that is chemically changed by light, thus stimulating an *action potential*. Photoreceptors are often grouped together to form specialised light-sensitive organs, such as eyes.

■ *e.g.* A *rod cell* in the *retina* of vertebrates is an example of a photoreceptor.

photosynthesis: the process by which *organic* compounds are synthesised from carbon dioxide and water using light energy.

■ Photosynthesis takes place in the *chloroplasts* of *photoautotrophs*. It occurs in two stages, known as the *light-dependent reaction* and the *light-independent reaction*. The chemical reactions that take place during photosynthesis can be summarised by the diagram below.

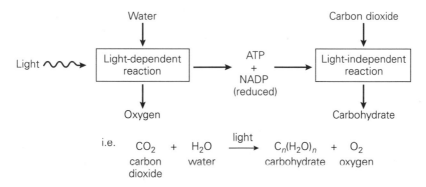

i.e.
$$CO_2 + H_2O \xrightarrow{light} C_n(H_2O)_n + O_2$$
carbon dioxide   water   carbohydrate   oxygen

■ *TIP* Remember that the light-dependent reaction can only take place in the light. The light-independent reaction requires substances produced in the light before it can proceed. Therefore the light-independent reaction also takes place in the light.

photosystem: a system of photosynthetic pigments found in *chloroplasts*.

■ Photosystems are found in the *thylakoid* membranes of chloroplasts and are involved in the *light-dependent reaction* of *photosynthesis*. Each photosystem contains about 300 molecules of *chlorophyll* that trap light energy, which is then passed on to a reaction centre (chlorophyll a molecule). In photosystem I (PSI), chlorophyll a molecules, known as P700, absorb light of wavelength 700 nm. Similarly, in PSII, chlorophyll a molecules, known as P680, absorb light of wavelength 680 nm. Within the light-dependent reaction, cyclic *photophosphorylation* involves PSI only, whereas non-cyclic photophosphorylation involves both PSI and PSII.

phototropism: a change in *growth* in response to light.

■ Most stems are positively phototropic because they grow towards light. Roots, on the other hand, are negatively phototrophic, growing away from light. The phototrophic response is thought to be regulated by *plant growth substances*.

phylum: a group used for the *classification* of organisms.

■ Each *kingdom* is divided into a number of phyla (plural), which are therefore the second largest groups used in classification.

■ *e.g.* Bryophyta, *Arthropoda* and *Chordata*.

physiology: the study of processes which sustain life in organisms, such as nutrition, *respiration* and reproduction.

phytochrome: a *pigment* found in plants which is important in regulating *growth*.

■ Phytochrome exists in two forms. P_R has a maximum light absorption peak in red light of wavelength 660 nm, whereas P_{FR} absorbs maximally in far-red light at 730 nm. The two forms are interconvertible, as shown in the diagram below.

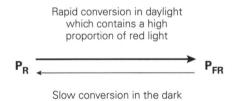

Rapid conversion in daylight which contains a high proportion of red light

$$P_R \longrightarrow P_{FR}$$

Slow conversion in the dark

P_{FR} is thought to be metabolically active and influences a number of light-related processes, including *germination* and flowering.

pigment: a coloured substance.

■ Pigments have a wide variety of functions in living organisms.

■ *e.g.* Chlorophyll, *rhodopsin* and *haemoglobin*.

pinocytosis: the process by which cells take up liquids from their environment.

■ Pinocytosis is an example of *endocytosis*. The cell surface membrane invaginates to capture the liquid and then encloses around it to form a *vacuole*.

■ *TIP* Try not to confuse this process with *phagocytosis*, which is the process by which cells actively take up solid particles from their environment.

pituitary gland: an *endocrine* gland situated at the base of the *brain*.

■ The pituitary gland is attached to the *hypothalamus* and has two lobes, anterior and posterior. The anterior lobe produces and secretes numerous *hormones* when stimulated by chemical releasing factors from the hypothalamus. These hormones include *thyroid-stimulating hormone (TSH), follicle-stimulating hormone (FSH), luteinising hormone (LH)* and adreno-corticotrophic hormone (ACTH). The posterior lobe does not produce hormones, but releases oxytocin and *anti-diuretic hormone (ADH)* which are produced by the hypothalamus and stored in the posterior lobe. This process is known as neurosecretion. The interaction between the hypothalamus and the pituitary gland is summarised by the diagram below.

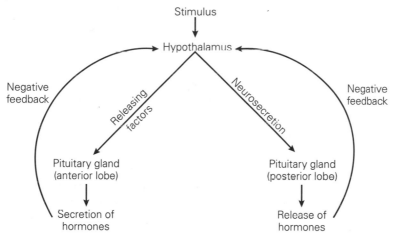

placenta: the *organ* which permits the exchange of materials between a mother and an *embryo*.

■ The placenta attaches the embryo to the wall of the *uterus* in mammals. It is composed of a combination of embryonic and maternal *tissues* and allows the exchange of materials between the circulatory systems of the embryo and the mother. It also has an important *endocrine* function, secreting a number of *hormones* such as *progesterone* and *oestrogen*. The placenta is expelled following birth.

Plantae: the *kingdom* containing plants.

■ Plants share the following features:
- they are multicellular organisms
- the cells possess *cell walls* composed mainly of *cellulose*
- they show *autotrophic nutrition*
- they have a life cycle which shows *alternation of generations*

■ *e.g.* Mosses (members of the *phylum Bryophyta*) and ferns (members of the phylum *Filicinophyta*).

plant growth substance: an *organic* compound that plays an important role in the coordination of plant growth and development.

■ *e.g.* Abscisic acid, auxin, cytokinin, gibberellin and ethene.

plasma: the liquid part of the blood.

■ Plasma is about 95% water and contains a number of dissolved substances:
- proteins, e.g. fibrinogen for *blood clotting*
- *inorganic ions,* e.g. Na^+, K^+, Cl^- and HCO_3^-
- nutrients, e.g. *glucose* and *amino acids*
- *hormones,* e.g. *insulin*
- *vitamins,* e.g. vitamin C
- excretory materials, e.g. *urea* and carbon dioxide

Plasma that has left the blood through capillary walls (and therefore does not contain the large *plasma proteins*) is known as *tissue fluid*. Plasma without the clotting factors is known as *serum*.

plasma membrane: the *cell membrane* that surrounds the cell.

▪ The plasma membrane is also known as the cell surface membrane.

plasma protein: any of the protein constituents of blood *plasma*.

▪ *e.g.* Fibrinogen.

plasmid: a small circular molecule of *DNA* found in *bacteria*.

▪ Plasmids sometimes contain important *genes*, such as those conferring resistance to *antibiotics*. They are useful as *vectors* in *gene technology*.

plasmodesmata: fine threads of *cytoplasm* that pass through pores in the *cell walls* of adjacent plant cells.

▪ Plasmodesmata are part of a continuous cytoplasmic system in plants. They allow substances such as mineral ions to move from one cell to the next without having to pass through cell walls and cell surface membranes, known as the *symplastic pathway*.

plasmolysis: the shrinkage of the *cytoplasm* away from the *cell wall* in plant cells.

▪ If a plant cell is placed in a solution with a lower (more negative) *water potential*, water will leave the cell by *osmosis* and the cytoplasm will shrink away from the cell wall. Plasmolysis may lead to *wilting* of the plant as the cells lose their normal turgidity.

platelets: microscopic cell fragments found in mammalian blood.

▪ Platelets (thrombocytes) play an important role in *blood clotting*.

Platyhelminthes: the animal *phylum* containing flatworms, tapeworms and flukes.

▪ Platyhelminthes share the following features:
  • they are flattened, unsegmented worms
  • they do not have a *coelom*
  • they have a branched gut with a single opening

plumule: the part of a plant *embryo* that develops into a shoot.

pollen: a collection of grains each containing a male *gamete* of a flowering plant.

▪ The generalised structure of a mature pollen grain is shown in the diagram below.

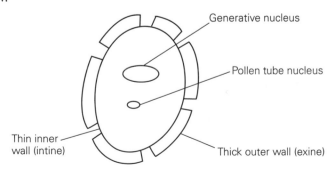

Generative nucleus

Pollen tube nucleus

Thin inner wall (intine)

Thick outer wall (exine)

The exact structure of pollen grains depends upon the *species* of plant in question. In general, pollen from insect-pollinated plants is textured and relatively large. Pollen from wind-pollinated plants is smaller and lighter, and usually has a smooth texture.

■ *TIP* Note that a pollen grain is not the male gamete of a flowering plant. It contains the male gamete.

pollen tube: a tube produced by a germinating pollen grain.

■ The pollen tube is used to transport the male *gamete* to the ovule in flowering plants.

pollination: the transfer of *pollen* from an *anther* to a *stigma* in flowering plants.

■ Self-pollination occurs if pollen is transferred from an anther to a stigma of the same plant. Cross-pollination is the transfer of pollen to the stigma of a different plant of the same *species*. This is usually achieved by the use of wind or insect *vectors*. There are several differences between wind-pollinated and insect-pollinated plants, as summarised by the table below.

Feature	Wind-pollinated	Insect-pollinated
Petals	Absent, or small and not colourful	Large and colourful
Nectar	Absent	Present
Stamens	Hang outside flower	Within flower
Stigmas	Feathery, outside flower	Sticky, inside flower
Pollen	Smaller grains, produced in large amounts, smooth and dry surface	Larger grains, produced in smaller amounts, sticky or rough surface

■ *TIP* Note that pollination is not the same as *fertilisation*, which involves the actual fusion of male and female *gametes*.

pollution: a change in the *abiotic* or *biotic* characteristics of the *environment* as a result of human activities.

■ *e.g.* Water pollution includes the leaching of *fertilisers* into rivers and lakes, leading to *eutrophication*. Air pollution includes carbon dioxide produced by the burning of fossil fuels which contributes to the *greenhouse effect*. Thermal pollution includes the release of hot water into rivers by power stations. This reduces the concentration of dissolved oxygen in the water and may lead to the death of many organisms.

■ *TIP* Some pollutants are biodegradable and may not cause damage if adequately dispersed or treated. Other pollutants are non-biodegradable and may accumulate in the environment and be concentrated in *food chains*.

polygenic inheritance: the inheritance of a character controlled by a number of *genes*.

■ Each gene in a polygenic system will have a small, but significant effect on the *phenotype*. Environmental factors may also influence the expression of polygenic characteristics, which usually show *continuous variation*.

■ *e.g.* Height in humans is a characteristic which shows polygenic inheritance.

polymer: a large molecule consisting of a number of smaller molecules.

■ The individual molecules within a polymer are known as *monomers*. These are usually joined together by *condensation* reactions in a process called polymerisation.

■ *e.g.* Starch is a polymer of *glucose*; *polypeptides* are polymers of *amino acids*; and *nucleic acids* are polymers of *nucleotides*.

polymerase chain reaction (PCR): a technique used to replicate a particular sequence of *DNA*.

■ The PCR is used as a means of increasing the quantity of genetic material available for sequencing a particular *gene* or for measuring gene expression. The basic processes involved in the PCR are summarised in the flow chart below.

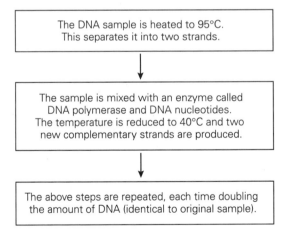

The DNA sample is heated to 95°C.
This separates it into two strands.

The sample is mixed with an enzyme called DNA polymerase and DNA nucleotides. The temperature is reduced to 40°C and two new complementary strands are produced.

The above steps are repeated, each time doubling the amount of DNA (identical to original sample).

polynucleotide: a long chain of *nucleotides*.

■ Polynucleotides usually consist of hundreds or thousands of nucleotides joined together by phosphodiester bonds.

■ *e.g.* DNA and RNA are polynucleotides.

polypeptide: a long chain of *amino acids*.

■ Polypeptides usually contain 30–300 amino acids, joined together by *peptide bonds*. The sequence of amino acids in a polypeptide is known as its *primary structure*. This determines its *secondary structure* and, in globular proteins, its *tertiary structure*. One or more polypeptides form a *protein*.

■ *e.g.* Adrenocorticotrophic hormone (ACTH) is a short polypeptide containing only 39 amino acids.

polyploidy: a condition in which an organism has three or more sets of *chromosomes* in each cell.

■ Most organisms have a *diploid* (2n) number of chromosomes in each cell. Polyploids, however, may be triploid (3n), tetraploid (4n), hexaploid (6n) and so on. There are two types of polyploidy — autopolyploidy and allopolyploidy. Autopolyploidy is known as 'self-polyploidy' and occurs when chromosomes fail to separate during cell division. All the chromosomes are from the same *species*. Allopolyploidy is known as 'cross-polyploidy' and involves hybridisation between two different species. The resulting hybrid is usually sterile (because *homologous chromosomes* cannot form pairs during *meiosis*), but fertility is restored when the chromosome number doubles by a mechanism similar to autopolyploidy. Polyploidy is a form of chromosome *mutation* and is much more common in plants than in animals.

■ *e.g.* Bramley cooking apples are triploid (and therefore sterile).

■ *TIP* Try not to confuse this term with *polysomy*, in which an organism has one more or one less chromosome than the normal complement. A common examination question asks you to distinguish these two terms.

polysaccharide: a long chain of *monosaccharides*.

■ Polysaccharides are *carbohydrates* which consist of large numbers of monosaccharides joined together by glycosidic bonds.

■ *e.g. Starch*, *glycogen* and *cellulose*.

polysomy: a condition in which an organism has one more *chromosome* than normal.

■ In cases of polysomy, the number of a particular chromosome is not *diploid*. In other words, there may be one or three copies of the chromosome rather than the expected two copies. Polysomy is usually caused by *non-disjunction* during *meiosis*, but may also be due to a translocation *mutation*.

■ *e.g. Down's syndrome* is an example of polysomy where affected individuals possess three copies (trisomy) of chromosome 21.

■ *TIP* Try not to confuse this term with *polyploidy*, in which an organism has three or more sets of chromosomes in each cell. A common examination question asks you to distinguish these two terms.

population: a group of individuals of the same *species* occupying a certain area.

■ Members of a population are capable of inter-breeding and therefore share the same *gene pool*.

population growth curve: a graph that shows the growth of a *population* of organisms over a period of time.

■ A population growth curve can be divided into four stages:
- lag phase — no increase in number as the organisms adapt to the environmental conditions
- log phase — rapid population growth due to abundant food supplies and little toxic waste
- stationary phase — population remains more-or-less constant

- decline phase — many organisms die due to lack of food and a build-up of toxic waste

positive feedback: a condition in which the deviation of a physiological factor from its optimum causes further deviation to occur.

▨ *e.g.* If the body temperature falls significantly below the optimum, the metabolic rate also falls. This means that less heat is produced and the body temperature will fall further. Ultimately, this process may lead to hypothermia and even death.

▨ *TIP* It is important to distinguish positive feedback from *negative feedback*, in which the deviation of a physiological factor from its optimum stimulates mechanisms which return the factor back to the optimum.

predator: an animal that kills and eats other animals.

▨ All predators are *consumers* and they feed on other consumers (prey).

▨ *e.g.* Lions (predators) feed on zebras (prey).

pressure potential: the pressure produced by the *cell wall* pushing against the cell surface membrane in plants.

▨ The relationship between *water potential* (ψ), *solute potential* (ψ_S) and pressure potential (ψ_P) can be summarised by the equation:

$$\psi = \psi_S + \psi_P$$

In a plasmolysed cell, ψ_P is zero because the cell wall is not in contact with the cell surface membrane. In *turgid cells*, ψ_P has a positive value, which will increase as the water content of the cell increases.

primary consumer: an organism that feeds on green plants.

▨ All primary consumers are *herbivores* and they feed on *producers*.

▨ *e.g.* Rabbits are primary consumers.

primary structure: the sequence of *amino acids* that makes up a *protein*.

▨ There are about 20 different types of amino acid and a typical protein contains between 100 and 600 amino acids. Variation in the number and sequence of amino acids in a molecule means that it is possible to have an enormous number of different proteins.

producer: an *autotrophic* organism that forms the first *trophic level* in a *food chain*.

▨ Most terrestrial food chains are based on green plants, which convert light energy to chemical energy by *photosynthesis*. Some of this energy is then transferred to *primary consumers* (*herbivores*) when the plants are eaten.

progesterone: a sex *hormone* produced by the *ovaries* in females.

▨ Progesterone is secreted by the *corpus luteum* and plays an important role in the regulation of the *menstrual cycle*. During the first few weeks of pregnancy, the corpus luteum continues to secrete progesterone, but in the later stages this is taken over by the *placenta*. Progesterone maintains the lining of the uterus for *implantation* and inhibits contraction of the uterine muscles.

▨ *TIP* Make sure that you know how progesterone interacts with *follicle-stimulating hormone (FSH)*, *oestrogen* and *luteinising hormone (LH)* to control the menstrual cycle.

Prokaryotae: the *kingdom* which includes *bacteria*.

■ Prokaryotic cells have a different structure to all other (*eukaryotic*) cells. The main differences are summarised in the table below.

Feature	Prokaryotic cells	Eukaryotic cells
Size	Cells smaller, usually less than 5 µm in diameter	Cells larger, often as much as 50 µm in diameter
Nucleus and DNA	Cells do not have a nucleus. The DNA, which is not associated with proteins, is present as a circular strand in the cytoplasm of the cell	Cells have a nucleus. The DNA, which is in long strands, is associated with proteins forming chromosomes
Organelles	Few organelles present and none of them are surrounded by a membrane	Many membrane-surrounded organelles, such as mitochondria, present
Ribosomes	Only have small ribosomes which are free in the cytoplasm	Have small ribosomes and larger ones which are associated with membranes forming rough endoplasmic reticulum

prolactin: a *hormone* that stimulates the secretion of milk in mammals.

■ Prolactin is secreted following birth by the anterior *pituitary gland* of the mother. This stimulates the synthesis of milk by the mammary glands. A second hormone, oxytocin, is required for the milk to be released to a suckling infant (see *lactation*).

protandry: the condition in which the male reproductive organs of a *flower* mature before the female reproductive organs.

■ Protandry ensures that self-fertilisation (which may be disadvantageous for the plant) does not occur.

■ *e.g.* Ivy has protandrous flowers.

■ *TIP* Try not to confuse this term with *protogyny*, in which the female reproductive organs mature before the male ones.

protease: any *enzyme* that catalyses the breakdown of proteins.

■ Proteases break down proteins into peptides and *amino acids* by *hydrolysis*. They can be classified into two categories, *endopeptidases* and *exopeptidases*, depending on which *peptide bonds* they break. Several proteases are normally required for the complete *digestion* of a protein.

■ *e.g.* Pepsin and *trypsin*.

protein: an *organic* molecule composed of long chains of *amino acids*.

■ The amino acids in a protein are joined together by *peptide bonds*. The long chains of amino acids (known as *polypeptides*) may have four different levels of structure:
- *primary structure* is the sequence of amino acids
- *secondary structure* is the coiling of a polypeptide into a helix

- *tertiary structure* is the folding of a polypeptide into a globular shape
- *quaternary structure* is the association of more than one polypeptide chain

Many of the functions of proteins can be explained in terms of their structure. Fibrous proteins are usually long, coiled strands or flat sheets and have a structural function, such as *collagen, actin, myosin* and *fibrin*. Globular proteins have a roughly spherical shape and have a physiological function, such as *enzymes, antibodies, haemoglobin* and *insulin*.

■ *TIP* Proteins are made up from amino acids and so contain the elements carbon, hydrogen, oxygen and nitrogen. They sometimes contain sulphur but never contain phosphorus.

protein synthesis: the synthesis of a *protein* from its constituent *amino acids*.

■ Protein synthesis takes place on *ribosomes* according to the sequence of bases in *DNA* that constitute a *gene*. The process of protein synthesis is summarised in the diagram below.

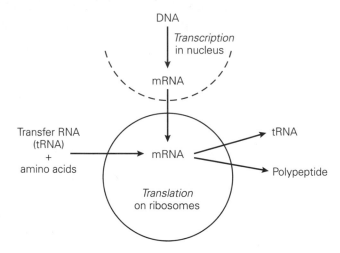

Protoctista: the *kingdom* containing *eukaryotic* organisms that cannot be classified as members of the *Fungi, Animalia* or *Plantae* kingdoms.

■ *e.g.* Red algae, slime moulds and *Amoeba*.

protogyny: the condition in which the female reproductive organs of a *flower* mature before the male reproductive organs.

■ Protogyny ensures that self-fertilisation (which may be disadvantageous for the plant) does not occur.

■ *e.g.* Some buttercup species have protogynous flowers.

■ *TIP* Try not to confuse this term with *protandry*, in which the male reproductive organs mature before the female ones.

proximal convoluted tubule: the part of a kidney *nephron* located between *Bowman's capsule* and the *loop of Henle*.

■ *Glucose, amino acids,* salts and water are selectively reabsorbed from the *glomerular filtrate* in the proximal convoluted tubule. Reabsorption is by *diffusion* and *active transport* and the cells lining the tubule have a large surface area and many *mitochondria* to facilitate these processes.

pulmonary artery: the blood vessel which carries blood from the *heart* to the *lungs*.

■ The pulmonary *artery* carries deoxygenated blood from the right *ventricle* of the heart to the lungs. It is the only artery to carry deoxygenated blood.

pulmonary vein: the blood vessel which carries blood from the *lungs* to the *heart*.

■ The pulmonary *vein* carries oxygenated blood from the lungs to the left *atrium* of the heart. It is the only vein to carry oxygenated blood.

purine: an *organic* base found in *nucleotides*.

■ The two purines found in nucleotides are *adenine (A)* and *guanine (G)*.

■ *TIP* Try not to confuse these molecules with *pyrimidines*, which are also organic bases found in nucleotides, but which have a different molecular structure and are smaller than purines.

Purkyne tissue: a specialised *tissue* which helps to coordinate muscle contraction in the mammalian *heart*.

■ The Purkyne tissue originates in the *bundle of His* and spreads out over the *ventricles* of the heart. Electrical activity generated by the *sino-atrial node* is conducted rapidly through the Purkyne tissue, ensuring that both ventricles contract simultaneously.

pyramid of numbers/biomass/energy: three different ways of describing the relationships between different *trophic levels* in a *food chain*. A pyramid of numbers shows the number of individuals at each trophic level in a food chain. A pyramid of biomass shows the mass of organisms at each trophic level in a food chain. A pyramid of energy shows the amount of energy available at each trophic level in a food chain.

■ The diagram below shows the shape of these pyramids for a simple food chain. Note that the area of the boxes represents the number, biomass or energy.

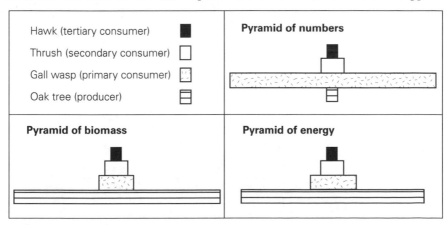

■ *TIP* Not all pyramids of numbers have the characteristic upright pyramid shape. The shape varies according to the size of the organisms involved.

pyrimidine: an *organic* base found in *nucleotides*.

■ The three pyrimidines found in nucleotides are *cytosine (C)*, *thymine (T)* and *uracil (U)*. *DNA* contains only C and T, whereas *RNA* contains only C and U.

■ *TIP* Try not to confuse these molecules with *purines*, which are also organic bases found in nucleotides, but which have a different molecular structure and are larger than pyrimidines.

pyruvate: the end product of *glycolysis*.

■ Pyruvate is produced from *glucose* during glycolysis according to the summary equation below.

glucose + 2NAD (oxidised) + 2ADP + 2Pi $\longrightarrow$ 2 × pyruvate + 2NAD (reduced) + 2ATP

In *aerobic respiration*, pyruvate is converted to *acetylcoenzyme A* in the *link reaction* and this compound enters the *Krebs cycle*. In *anaerobic respiration*, pyruvate is converted to lactic acid (animals) or ethanol and carbon dioxide (plants and microorganisms).

Q_{10}: the increase in the rate of a process when the temperature is increased by 10°C.

■ Q_{10} is known as the temperature coefficient and can be calculated from the simple formula given below.

$$Q_{10} = \frac{\text{rate of reaction at } (T + 10)°C}{\text{rate of reaction at } T°C}$$

■ **e.g.** At temperatures between 0°C and approximately 40°C, most *enzyme*-controlled reactions will have $Q_{10} = 2$. This means that a ten degree rise in temperature will cause the rate of an enzyme-controlled reaction to double.

quadrat: a square frame used to mark out an area in which organisms can be investigated.

■ Quadrats are usually a square metre in area and may be divided into a grid by cross wires. They are sometimes placed at random in a given area in order to sample the organisms present.

■ **TIP** Note that throwing a quadrat over your shoulder will not result in random distribution. The only way this can be done is to use random numbers to identify coordinates on a grid of the area and use the quadrat at these points.

quaternary structure: the type of structure shown by a *protein* made up from more than one *polypeptide* chain.

■ The quaternary structure of a protein specifies the structural relationship of the component polypeptides.

■ **e.g.** *Haemoglobin* has a quaternary structure consisting of four polypeptide chains associated with four *haem* groups.

radial symmetry: describes an organism that can be divided in any longitudinal plane to give two approximately identical halves.

▓ Members of the phyla *Cnidaria* and *Echinodermata* exhibit radial symmetry. Most other animals are *bilaterally symmetrical* and can be divided in only one plane to give two identical halves.

radicle: the part of a plant *embryo* that develops into a *root*.

receptor: a cell or group of cells specialised to detect a particular *stimulus*.

▓ If the stimulus exceeds a minimum threshold value, the receptor will be depolarised and an *action potential* is set up in a sensory *neurone*. Receptors are often grouped together to form sense organs.

▓ *e.g.* Baroreceptors (blood pressure), chemoreceptors (blood *pH*) and *rods* and *cones* (light) in the *retina* of the eye.

▓ *TIP* The term receptor may also be used to refer to a specific *protein* molecule found on the cell surface membrane of a cell. A *hormone* or *neurotransmitter* may bind to such a receptor and trigger a physiological response by the cell.

recessive: any *allele* that is only expressed in the *phenotype* of an organism if the *dominant* allele is not present.

▓ *e.g.* In peas, the allele for a short plant (t) is recessive to that for a tall plant (T). Therefore, only pea plants with the *genotype* tt will be short.

▓ *TIP* Alleles may be recessive, *dominant*, or *codominant*. It is important that you know and understand all of these terms. In addition, just because an allele is dominant, it does not mean that organisms with this allele will be more healthy or that the allele will eventually be found in every member of the *species*. For example, a number of dominant alleles found in humans have lethal effects!

recombinant: an organism containing a combination of *alleles* not present in the parents.

▓ Recombinants arise through three processes:
- *crossing-over* between *homologous chromosomes* during *meiosis*
- independent assortment of chromosomes during meiosis
- random fusion of male and female *gametes* during *fertilisation*

red blood cell: a cell that is specialised for transporting respiratory gases in the blood.

■ Red blood cells are also known as erythrocytes. Each cell is about 8 µm in diameter and has a biconcave shape, providing a high surface area to volume ratio for efficient gas exchange. Mammalian red blood cells have no *nucleus*, enabling them to carry large amounts of the oxygen-carrying pigment *haemoglobin*.

reducing sugar: any *carbohydrate* which has reducing properties.

■ All *monosaccharides* and most *disaccharides* are reducing sugars (they can donate *electrons* to other molecules). Reducing sugars produce a positive result with *Benedict's test*.

■ *e.g.* Glucose, *maltose* and *lactose*.

■ *TIP* Remember that *sucrose* is not a reducing sugar.

reduction: the addition of an *electron* to a molecule.

■ Reduction is also used to describe the addition of hydrogen or the removal of oxygen from a molecule. The reduction of one molecule is always accompanied by the *oxidation* of another. In biological systems, reduction and oxidation are controlled by *enzymes* called *oxidoreductases*.

■ *TIP* Make sure that you do not confuse reduction and oxidation, which is the loss of an electron by a molecule. Remember OILRIG (Oxidation Is Loss Reduction Is Gain).

reflex: a rapid automatic response to a particular *stimulus*.

■ A reflex response is mediated by a simple nervous circuit called a reflex arc, as shown in the diagram below.

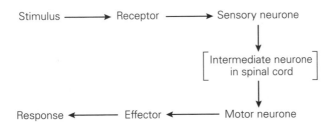

■ *e.g.* If a hand is accidentally placed on a hot object, it is quickly pulled away. Reflexes are also important in responding to internal changes, such as the need for more oxygen in the blood, by increasing the breathing rate.

refractory period: the time interval during which a nerve cell is incapable of responding to a *stimulus*.

■ After a nerve cell has responded to a stimulus, certain ionic movement must occur and the *resting potential* has to be restored. During these processes, the nerve cell is incapable of responding to a second stimulus. A typical refractory period lasts for approximately 3 ms and it limits the number of impulses that can be transmitted along a nerve cell in a given period of time.

renal capsule: see *Bowman's capsule*.

renin: an *enzyme* secreted by the *kidney* in response to a fall in blood pressure.

■ Renin stimulates the secretion of *aldosterone* by the *adrenal glands*. Aldosterone, in turn, stimulates the reabsorption of sodium ions and water in the kidney, thus raising blood pressure.

reproductive isolation: the prevention of interbreeding between members of different *species* that inhabit the same geographical area.

■ There are two main types of reproductive isolating mechanisms.

- Pre-zygotic isolating mechanisms: which prevent *fertilisation* taking place, such as seasonal isolation (the breeding seasons of the different species do not overlap) and behavioural isolation (different courtship behaviour ensures that mating only takes place between members of the same species).

- Post-zygotic isolating mechanisms: the offspring produced by interbreeding between two species either do not survive to breed (hybrid inviability) or are sterile (hybrid infertility).

If two populations of the same species living in the same area become reproductively isolated, this may lead to *sympatric speciation*.

■ *TIP* Note that geographical isolation will also prevent interbreeding between two populations of the same species, which may lead to *allopatric speciation*.

resistance: the ability of an organism to tolerate substances which would normally be toxic to them.

■ Resistant organisms usually carry a *gene* for tolerance to a particular substance. These genes may arise as a result of *mutation* and are often found in very low frequencies in areas where toxic chemicals are not used. The use of toxic chemicals will result in the death of organisms not possessing the gene and therefore select for those that are resistant. The resistant organisms will then interbreed, passing on their genes, and eventually the whole population may become resistant.

■ *e.g.* Certain strains of *bacteria* are resistant to *antibiotics*; mosquitoes develop strains resistant to insecticides; and rats have developed resistance to the poison warfarin.

■ *TIP* Note that the use of toxic chemicals does not cause resistance (a small proportion of the population was already resistant), it simply selects those organisms which are resistant to the chemical. It is better to avoid using the term 'tolerance', as this usually refers to a gradual adjustment that takes place during the lifetime of an organism. Finally, remember that resistance is not the same as immunity.

respiration: the *oxidation* of *organic* molecules by cells in order to release energy.

■ The main respiratory substrate is *glucose*, which is oxidised in cells to produce *ATP*. *Aerobic respiration* involves *glycolysis*, the *link reaction*, the *Krebs cycle* and the *electron transport chain*. The process of aerobic respiration can be summarised by the equation below.

$$C_6H_{12}O_6 + 6O_2 \longrightarrow 6CO_2 + 6H_2O + 34\text{–}38 \text{ molecules ATP}$$

glucose   oxygen   carbon   water   energy
dioxide

Anaerobic respiration involves only glycolysis and results in a much lower yield of ATP.

■ *TIP* Note that respiration is a cellular process and should not be used to refer to gas exchange with the environment.

respiratory quotient (RQ): the ratio of the volume of carbon dioxide produced to the volume of oxygen consumed during *respiration*.

■ The theoretical RQ for respiring *carbohydrates* is 1.0, for *lipids* is 0.7 and for *proteins* is 0.9. However, interpretation of the RQ is often complex, as an organism may respire a mixture of substrates at the same time. Furthermore, the use or production of carbon dioxide by other metabolic processes will also affect the respiratory quotient.

resting potential: the difference in electrical charge across the cell surface membrane of a nerve cell.

■ The inside of a nerve cell at rest is negatively charged (approximately –60 mV) relative to the outside, due to the *active transport* of positive sodium ions out of the cell. This difference is important as it allows the generation of an *action potential* following stimulation of the cell.

restriction endonuclease: an *enzyme* that cuts *DNA* molecules at specific sites.

■ Restriction endonucleases are produced by many *bacteria* and are sometimes referred to simply as restriction enzymes. The cleavage site of a particular restriction endonuclease is related to a specific sequence of bases. Two examples of such sequences are shown below.

G T T ¦ A A C
C A A ¦ T T G

The enzyme *Hpa1* recognises this base sequence and cuts it as shown by the dotted line

G¦T T A A C
C A A T T¦G

The enzyme *EcoR1* recognises the same sequence but cuts it in a different way

Restriction endonucleases have a number of uses in *gene technology*.

retina: a layer of light-sensitive cells found at the back of the eye.

■ There are two types of light-sensitive cell in the retina: *rod cells* and *cone cells*. The differences between these two cell-types are summarised in the table below.

Feature	Rods	Cones
Location	Throughout the retina (except in the fovea)	Concentrated in the fovea
Sensitivity to light	Higher	Lower
Colour vision	No	Yes
Visual acuity	Low	High

Information from the rods and cones is processed in the retina before being sent via the optic nerve to the *brain*.

retrovirus: a type of *virus* that contains *RNA* as its *nucleic acid*.

Retroviruses are capable of converting their RNA into *DNA* using an *enzyme* known as reverse transcriptase. This enables the viral DNA to be integrated into the host DNA.

e.g. Human immunodeficiency virus (HIV) is an example of a retrovirus.

Rhizobium: a genus of nitrogen-fixing *bacteria* which form a mutualistic relationship with *leguminous plants*.

Rhizobium lives in the root nodules of leguminous plants and fixes atmospheric nitrogen (N_2) to ammonia (NH_3). The plants are able to use this source of ammonia to produce *amino acids* and *nucleic acids*, while the bacteria receive *carbohydrate* from the plant. As both *species* benefit from the relationship, this is an example of *mutualism*.

rhizoid: hair-like structure that acts as a simple *root* in certain plants.

Rhizoids are found in mosses and liverworts, where they anchor the plant into the ground and aid the absorption of water and mineral salts.

rhodopsin: a light-sensitive *pigment* found in the *rod cells* of the *retina*.

Rhodopsin consists of a *protein*, opsin, linked to retinine (a derivative of *vitamin* A). The presence of light causes rhodopsin to split into its two components, initiating the transmission of a nerve *impulse* along the optic nerve to the *brain*. Before it can be stimulated again, rhodopsin must be resynthesised from opsin and retinine. This is an *anabolic* reaction and requires *ATP*.

ribonucleic acid: see *RNA*.

ribose: a five-carbon sugar (*pentose*) found in *RNA*.

TIP Try not to confuse this molecule with deoxyribose, which is a five-carbon sugar found in *DNA*. Deoxyribose has one less oxygen atom per molecule than ribose.

ribosome: a small *organelle* that acts as a site of *protein synthesis* in a cell.

Ribosomes are spherical structures composed of ribosomal *RNA* and protein. They are found free in the *cytoplasm* and attached to *endoplasmic reticulum* (making it rough endoplasmic reticulum).

ribulose bisphosphate (RuBP): a five-carbon sugar which combines with carbon dioxide during the *light-independent reaction* of *photosynthesis*.

Ribulose bisphosphate combines with carbon dioxide to form two molecules of glycerate 3-phosphate. This is then reduced to triose phosphate, which can be used to produce *glucose* and *starch*, or to resynthesise ribulose bisphosphate via the *Calvin cycle*.

RNA (ribonucleic acid): a type of *nucleic acid* molecule.

There are three forms of RNA: *messenger RNA (mRNA)*, *transfer RNA (tRNA)* and ribosomal RNA (rRNA). All three forms play an important role in *protein synthesis*. The structure of RNA is similar to that of *DNA*, in that they are both *polynucleotides.* However, there are several key differences between these two types of nucleic acid, as shown in the table below.

Feature	DNA	RNA
Sugar	Deoxyribose	Ribose
Bases	Adenine, cytosine, guanine, thymine	Adenine, cytosine, guanine, uracil
Number of chains	2	1
Number of types	1	3 (mRNA; tRNA; rRNA)

In addition to its role in protein synthesis, RNA is the hereditary material in *retroviruses*.

RNA polymerase: an *enzyme* that catalyses the synthesis of *RNA*.

■ RNA polymerase synthesises RNA from other RNA or from *DNA*. Different types of RNA polymerase are used to produce the three different types of RNA: *messenger RNA, transfer RNA* and ribosomal RNA.

■ *e.g.* RNA polymerase II is used to synthesise messenger RNA from DNA during *transcription.*

rod cell: a light-sensitive cell found in the *retina* of the eye.

■ Rod cells contain the pigment *rhodopsin.* They are responsible for black-and-white vision and function effectively even in very dim light. They are found throughout the retina, except in one area which is packed with *cone cells* (the fovea).

■ *TIP* A common examination question asks you to compare rod cells with cone cells in the retina.

root: the part of a plant which holds it in the soil and absorbs water and mineral salts.

■ Most roots grow below the ground and show adaptations for their dual functions of anchorage and uptake of water and salts. The *vascular* tissue (*xylem* and *phloem*) is arranged at the centre of the root in order to combine strength and flexibility. The roots also provide a large surface area for the efficient absorption of water and salts. Some roots are important storage organs and are commercially important. These include carrots, turnips and sugar beet.

root pressure: the force exerted by *roots* which pushes water into the *xylem* and up the plant.

■ It is thought that root pressure is due to both the *osmosis* of water from the soil into the root cells, and the *active transport* of salts into the xylem. These processes maintain a *water potential* gradient along which the water will move up the plant.

ruminant: any mammal that possesses a rumen.

■ A rumen is a chamber found at the anterior end of the gut. It is an adaptation found in mammals which feed predominantly on grass. Microorganisms in the rumen digest *cellulose* in the grass and then the mixture (known as the 'cud') is brought back to the mouth for further chewing. The cud is then swallowed and enters the digestive system as normal. This process allows ruminants to use the *glucose* formed from the breakdown of cellulose by the resident micro-organisms.

saliva: a watery fluid secreted by the salivary glands in the mouth.

■ The production of saliva is stimulated by the thought, sight or smell of food, or by the presence of food in the mouth. Saliva contains *mucus* to lubricate the passage of food and salivary *amylase*, which is an *enzyme* that begins the digestion of *starch*. The composition of saliva varies between *species*. For example, blood-sucking animals such as leeches have an anticoagulant in their saliva.

saltatory conduction: the mechanism by which an *impulse* is transmitted along a myelinated nerve cell.

■ Many nerve cells in mammals possess a *myelin* sheath. However, there are small gaps in this sheath, known as nodes of Ranvier. Only at these nodes can ions flow across the nerve *cell membrane*, propagating an *action potential*. As a result, impulses travel along nerve cells in a series of jumps, from one node to the next. This is known as saltatory conduction and it leads to impulses moving much more rapidly than in non-myelinated nerve cells.

sampling: the selection of a number of items or individuals from a *population*.

■ The advantages of sampling (compared with studying every item or individual in the population) are that it saves time and produces a manageable quantity of data for analysis. Furthermore, if a sample is chosen carefully, it will be representative of the whole population. In order to choose a sample that is free from bias, it is important to sample at random and to have a sample size that is as large as possible.

SAN: see *sino-atrial node*.

saprophyte: any organism that obtains its nutrients from dead or decaying matter by secreting digestive *enzymes* onto the food and then absorbing the soluble products of *digestion*.

■ Saprophytes play an important role in *nutrient recycling*, because they break down *organic* molecules and release *inorganic* molecules, such as carbon dioxide and ammonia.

■ *e.g.* Most *bacteria* and *fungi* are saprophytes.

sarcolemma: the membrane that surrounds a muscle fibre.

saturated fatty acid: a *fatty acid* which has the maximum number of hydrogen atoms attached to its carbon atoms.

■ Saturated fatty acids have no double bonds between their carbon atoms.

■ *e.g.* Stearic acid, $CH_3(CH_2)_{16}COOH$.

■ *TIP* Try not to confuse saturated fatty acids with *unsaturated fatty acids*, which have at least one double bond between their carbon atoms.

Schwann cell: a cell that produces the *myelin* sheath around a nerve cell.

■ Each Schwann cell is responsible for a separate section of the myelin sheath and these are separated from each other by small gaps known as nodes of Ranvier. These gaps are important for *saltatory conduction*.

sclerenchyma: a support *tissue* in plants.

■ Sclerenchyma cells have very thick *cell walls*, due to the deposition of *lignin* and additional *cellulose*. This means that the mature cells are dead (since lignin makes the cell wall impermeable to water and gases) and the tissue has considerable mechanical strength. Sclerenchyma is abundant in stems and in the midrib of leaves, where it has an important support role.

second messenger: a molecule found within a cell which is activated by the binding of a chemical messenger to a *receptor* on the cell surface membrane.

■ Many chemical messengers, such as *hormones* or *neurotransmitters*, do not enter their target cells. Instead they bind to receptors on the cell surface membrane and activate second messengers inside the cell. The second messenger then initiates a physiological response to the original *stimulus*, usually by activating a particular *enzyme*.

■ *e.g.* Cyclic AMP (cAMP) is a second messenger for many hormones. The role of cAMP is summarised in the diagram below.

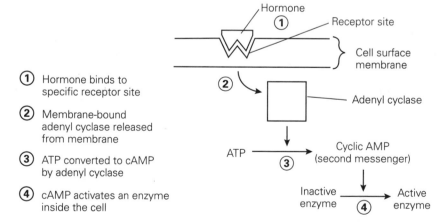

secondary structure: the way in which a *polypeptide* is coiled or folded.

■ The secondary structure of a polypeptide is determined by its *primary structure* (the sequence of *amino acids*). The polypeptide may be coiled, e.g. into an alpha-helix, or pleated, e.g. into a beta-sheet. Many fibrous *proteins* have this type of structure, such as *collagen* and keratin. Globular proteins show further folding into a *tertiary structure*.

seed: a structure containing the *embryo* of a flowering plant or conifer.

■ A seed develops from a fertilised ovule. In addition to the embryo, a seed usually contains a food store and has a protective coat known as the *testa*. The food store is either *endosperm* tissue (e.g. pine seed) or nutrients stored in the *cotyledons* (e.g. broad bean seed). Seeds are dispersed by one of a number of mechanisms and then usually lie dormant for a period of time. When conditions are favourable, the embryo plant begins to grow. This is known as *germination*.

seed bank: a collection of seeds which is used as a store of genetic material.

■ Seeds from a large number of plant *species* are kept in seed banks in conditions that keep them alive for many years. This is an example of *conservation*. It ensures that the *genes* contained within the seed are not lost and that the plant species does not become extinct. Unfortunately, the seeds of several species cannot be stored in this manner and whole plants or tissue cultures must be used to conserve these genes.

selection: the mechanism responsible for the *evolution* of new *species*.

■ The process of *natural selection* can be summarised as follows.

- All species tend to produce many more offspring than can ever survive. However, the size of populations tends to remain more or less constant, meaning that most of these offspring must die. It also follows that there must be competition for resources such as mates, food and territories. Therefore, there is a 'struggle for existence' among individuals.
- Individuals within a species differ from one another (*variation*) and many of these differences are inherited (genetic variation).
- Competition for resources, together with variation between individuals, means that certain members of the population are more likely to survive and reproduce than others. These individuals will inherit the characteristics of their parents and evolutionary change will take place through natural selection.

In addition to natural selection, species also evolve through sexual selection. This is the selection of any characteristic that contributes to reproductive success, regardless of whether it also aids survival. For example, the large colourful tail of the peacock makes it more likely to be eaten by predators. However, it also makes the peacock more attractive to females, leading to greater numbers of offspring. Therefore the large tail is selected despite its negative effect on survival.

■ *TIP* Natural selection exists in three forms — *directional selection, disruptive selection* and *stabilising selection*. It is useful to know the differences between these forms of selection.

selective breeding: the breeding of plants or animals in order to select certain favourable characteristics within the *population*.

■ Selective breeding uses artificial selection (selection by humans rather than by the *environment*) in order to establish a population which exhibits a commercially useful characteristic. All modern breeds of domestic animals, vegetables and fruits have been produced by selective breeding.

■ *e.g.* Friesian cows have been selectively bred in order to increase milk yields.

semi-conservative replication: the mechanism by which *DNA* produces identical copies of itself.

In semi-conservative replication, a molecule of DNA unwinds and each of the resulting strands acts as a template for the formation of a new strand. Therefore, the two daughter strands of DNA each consist of one strand of newly synthesised DNA and one strand of parental DNA. This process is summarised in the diagram below.

```
        ┌T A┐
        ├A T┤
        ├T A┤
        ├C G┤
        ├G  C┤
        └A   T┘
```
The polynucleotide strands of DNA separate

```
┌T A┐      ┌T A┐
├A T┤      ├A T┤
├T A┤      ├T A┤
├C G┤      ├C G┤
├G C┤      ├G C┤
├A  T┤      ├A  T┤
```
Each strand acts as a template for the formation of a new molecule of DNA

Individual nucleotides line up with complementary bases on parent DNA strands

```
┌T A┐      ┌T A┐
├A T┤      ├A T┤
├T A┤      ├T A┤
├C G┤      ├C G┤
├G C┤      ├G C┤
├A T┤      ├A T┤
```
The nucleotides are joined together by DNA polymerase to form two molecules of DNA. Each molecule contains a strand from the parent DNA and a new strand

semilunar valve: a valve that prevents the backflow of blood.

The semilunar valve in the *pulmonary artery* prevents blood flowing back into the right *ventricle* during relaxation of the *heart* and the valve in the *aorta* prevents backflow into the left ventricle. Semilunar valves also prevent blood flowing backwards in *veins*.

TIP Try not to confuse these valves with the *atrioventricular valves*, which are found between the atria and ventricles in the heart.

sepal: part of a *flower* that makes up the calyx.

Sepals are modified leaves and are usually green in colour. They form a protective layer around the flower when it is in the bud.

serum: blood *plasma* from which the fibrinogen and other clotting factors have been removed.

sex chromosomes: a pair of *chromosomes* which determine the sex of an organism.

Sex (*male* or *female*) is often determined by *genes* carried on the sex chromosomes. Female mammals have two *X chromosomes* and are known as the homogametic sex. Male mammals have one X and one *Y chromosome* and are known as the heterogametic sex.

sex determination: the mechanism by which sex is determined in a *species*.

■ Female mammals have two *X chromosomes* and are known as the homogametic sex. This is because all eggs (*gametes*) produced by a female contain an X chromosome. Male mammals, on the other hand, have one X and one *Y chromosome* and are known as the heterogametic sex because *sperm* may contain either an X chromosome or a Y chromosome. Since there is an equal chance that the sperm which fuses with an egg during *fertilisation* will have an X or a Y chromosome, it would be expected that approximately equal numbers of males and females will be produced.

■ *TIP* Note that in birds it is the male that is the homogametic sex and the female is the heterogametic sex. Other organisms may have different methods of determining sex, which may not be genetic. In crocodiles, for example, it is the temperature at which the eggs are incubated that determines whether the young will be male or female.

sex linkage: the inheritance of a characteristic which is determined by a *gene* on one of the *sex chromosomes*.

■ In most animals, the *Y chromosome* carries very few genes other than those determining sex. Sex-linked genes are, therefore, most likely to be found on the *X chromosome*.

■ *e.g.* Haemophilia is a sex-linked disorder characterised by very slow *blood clotting*, which is much more common in men than in women. This is because some of the genes governing blood clotting occur on the X chromosome. Women have two X chromosomes and so if one carries the *recessive* abnormal *allele* (h), there is a chance that its effects will be masked by a normal *dominant* allele (H) on the other X chromosome. Men, however, only have one X chromosome and so any abnormal alleles will be expressed. The *genotype* of a man with haemophilia is X^hY and the genotype of a normal (carrier) female is X^HX^h. The diagram (right) shows the possible genotypes and *phenotypes* of the offspring resulting from a cross between these two individuals.

■ *TIP* Note that when constructing a genetic cross involving a sex-linked trait, it is important to specify the sex of the individual using XX (female) or XY (male). (For further tips on how to do genetic crosses, see *monohybrid inheritance*.)

Parents:	Haemophiliac male	×	Normal (carrier) female
	X^hY		X^HX^h

Gametes: X^h Y X^H X^h

Offspring:

		Male gametes	
		X^h	Y
Female gametes	X^H	X^HX^h normal (carrier) female	X^HY normal male
	X^h	X^hX^h haemophiliac female	X^hY haemophiliac male

sexual reproduction: the fusion of male and female *gametes* to produce a *zygote*.

▨ In sexually reproducing organisms, the gametes of an individual are *haploid* (contain half the normal number of *chromosomes*) and are genetically different from each other due to *crossing-over* and the independent assortment of chromosomes during *meiosis*. Since male and female gametes fuse at random during *fertilisation* (and the male and female parents have different *genotypes*), there will be genetic *variation* in the *diploid* zygotes produced by sexual reproduction. This variation is important in *evolution*.

▨ *TIP* Make sure that you understand the differences between sexual reproduction and *asexual reproduction*, which is reproduction involving the formation of new individuals from a single parent without the fusion of gametes. It is possible to have sexual reproduction with just one parent — as seen in self-pollination of flowering plants.

single circulation: a type of circulatory system in which the blood passes only once through the *heart* in each complete circuit of the body.

▨ Single circulation can be seen in fish, as shown in the diagram below.

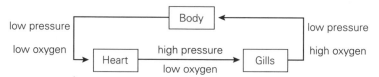

After blood has passed through the *gills*, it is oxygenated but at low pressure. A single circulation system is not as efficient, therefore, as a *double circulation* system (in which the blood passes twice through the heart in each complete circuit of the body).

sink: part of an organism where substances are removed from a *mass transport* system.

▨ Mass flow (transport) systems include the blood system of animals and the *phloem* in flowering plants. Substances are taken up or produced in one place (the *source*) and transported to another where they are used (the sink).

▨ *e.g.* Sucrose produced by *photosynthesis* in leaves (the source) is transported in the phloem to developing *fruits* (the sink).

sino-atrial node (SAN): a collection of specialised muscle cells in the right *atrium* of the *heart*.

▨ The SAN (pacemaker) initiates and maintains the *cardiac cycle*. Its activity can be regulated by the *autonomic nervous system*, which speeds up or slows down the cardiac cycle according to the body's requirements.

sinusoid: a blood-filled space in the *liver*.

▨ Blood enters the liver through the hepatic artery and the *hepatic portal vein*. It then flows from branches of these vessels through the sinusoids to branches of the hepatic vein. Sinusoids therefore replace *capillaries* in the liver, allowing direct contact between the blood and the liver cells (making the exchange of materials more efficient).

skeletal muscle: muscle tissue that is found attached to the *skeleton.*

■ Skeletal muscle is also known as voluntary muscle because it is mainly under conscious control. An individual muscle consists of a collection of muscle fibres (made up of *myofibrils*) each bounded by a *sarcolemma*. The muscle is attached at each end to a *bone* by *tendons*. Skeletal muscles are therefore responsible for the movement of joints and bones.

skeleton: the structure in an animal that provides mechanical support for the body, protection of internal *organs* and a framework for the attachment of *skeletal muscle.*

■ There are two basic types of skeleton — endoskeletons and exoskeletons. *Endoskeletons* are found inside the body of vertebrates and are composed of *bone* and *cartilage.* (Note that soft-bodied animals such as worms have a different type of internal skeleton, known as a *hydrostatic skeleton.*) *Exoskeletons* are found on the outer surface of arthropods and are composed mainly of *chitin.*

sliding-filament hypothesis: a hypothesis to account for the mechanism of contraction in *skeletal muscle.*

■ Skeletal muscle is composed of bundles of muscle fibres, which themselves are made up of a large number of *myofibrils*. The generalised structure of a myofibril is shown in the diagram below.

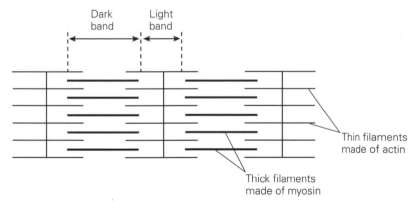

Upon stimulation, calcium ions bind to troponin molecules, causing the myosin-binding sites on the actin filament to become exposed. Cross-bridges are formed between actin and myosin, and the actin filaments are pulled together using energy from *ATP.* The process is then repeated (assuming the stimulation is still present) until the myofibril is fully contracted.

smooth muscle: muscle tissue that is not under conscious control.

■ Smooth muscle is also known as involuntary muscle and is controlled by the *autonomic nervous system.* It is found lining the digestive tract, the respiratory tract and many blood vessels.

solute potential: the component of *water potential* that is due to the presence of solute molecules.

■ The relationship between water potential (ψ), solute potential (ψ_S) and *pressure potential* (ψ_P) can be summarised by the equation:

$$\psi = \psi_S + \psi_P$$

Solute potential (ψ_S) always has a negative value as solutes lower the water potential of a solution.

source: part of an organism where substances enter a *mass transport system*.

■ Mass flow (transport) systems include the blood system of animals and the *phloem* in flowering plants. Substances are taken up or produced in one place (the source) and transported to another where they are used (the *sink*).

■ *e.g.* Sucrose produced by *photosynthesis* in leaves (the source) is transported in the phloem to developing *fruits* (the sink).

speciation: the formation of a new *species* of organisms.

■ Intraspecific speciation occurs when a single species gives rise to one or more new species. This requires *isolating* mechanisms to prevent gene flow between populations of the parent species. *Natural selection* may then act upon the isolated populations, leading to the development of new species. *Geographical isolation* will lead to *allopatric speciation* and *reproductive isolation* will lead to *sympatric speciation*. New species may also arise from the interbreeding of two different species (known as interspecific speciation). This is more common in plants than in animals, as it usually leads to an infertile hybrid which only regains its fertility by a doubling of the *chromosome* number (see *polyploidy*).

species: a group of organisms which are capable of interbreeding to produce fertile offspring.

■ A species is the smallest group used in *classification*. Each species occupies a particular ecological *niche*.

sperm: a *gamete* produced by male animals.

■ Sperm are produced in the *testes* by *spermatogenesis*. A single sperm cell consists of a head (containing the *haploid* nucleus), an *acrosome* for egg penetration and a *flagellum* (tail) for motility.

spermatogenesis: the production of *sperm* by the *testis*.

■ Certain cells (spermatogonia) within the testes divide by *mitosis* to produce large numbers of potential gametes called primary spermatocytes. The primary spermatocytes then undergo *meiosis* to produce sperm. Spermatogenesis is stimulated by *follicle-stimulating hormone (FSH)*.

sphincter muscle: a specialised muscle surrounding an opening within the body.

■ Relaxation and contraction of the muscle will open and close the orifice.

■ *e.g.* The pyloric sphincter is found at the lower end of the *stomach*, leading to the *duodenum*; small sphincter muscles are found at the end of *arterioles*, where they regulate blood flow to an organ.

spinal cord: part of the *central nervous system* that is enclosed by the backbone.

■ The spinal cord consists of a central cavity containing cerebrospinal fluid, surrounded by a core of grey matter (non-myelinated *neurones*) and an outer layer of white matter (myelinated neurones). The white matter contains numerous longitudinal neurones which conduct *impulses* to and from the *brain*.

The spinal cord also plays an important part in coordinating many *reflexes*.

spiracles: pores in the *exoskeleton* of an insect which lead to the *tracheae*.

■ The spiracles and tracheae form part of the gas exchange system of insects. Entry of oxygen into the spiracles is by *diffusion* at rest, but is increased by various ventilation mechanisms when activity is increased. Furthermore, the spiracles of most insects have a mechanism that allows them to open and close. This permits efficient gas exchange but prevents excessive water loss.

spore: a reproductive cell which gives rise to a new individual.

■ Spores are produced in large numbers by plants, *fungi, bacteria* and some *protoctists*, where they are primarily used for *asexual reproduction*. Spores often lie dormant during unfavourable conditions, before developing into a new individual. In plants which show *alternation of generations*, spores are formed by the *sporophyte* generation and give rise to the *gametophyte* generation.

■ *TIP* Try not to confuse spores with *gametes*, which have to fuse with other gametes (during *sexual reproduction*) before a new individual can develop.

sporophyte: the stage in the life cycle of a plant in which *spores* are produced.

■ The sporophyte stage alternates with the *gametophyte* stage (see *alternation of generations*). Sporophytes are *diploid* (containing two sets of *chromosomes*) and produce spores which develop into gametophytes. In mosses and liverworts, the gametophyte is the dominant stage. In flowering plants, however, the sporophyte is dominant.

stabilising selection: a type of *natural selection* in which the environment acts against forms at the extremes of the range of phenotypic variation.

■ The effect of stabilising selection is to favour phenotypes in the middle of the range, as shown in the diagram below.

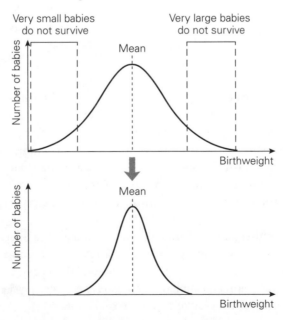

139

■ *e.g.* Babies whose birthweight is significantly below or above the mean of 3.6 kg have greater mortality than babies of average birthweight.

■ *TIP* Make sure that you understand the differences between stabilising selection and the other two forms of natural selection: *directional selection* and *disruptive selection*. Note that stabilising selection is the opposite of disruptive selection.

stamen: one of the male reproductive organs of a *flower*.

■ Each stamen consists of a filament (stalk) and an *anther*, which produces *pollen*.

standard deviation: a measure of the spread of data about the *mean*.

■ In addition to knowing the mean of a set of data, it is also useful to know the variability of the data. This can be described simply by looking at the difference between the largest and smallest values (the range). However, the standard deviation is a more useful measure of dispersion because it uses all the data to calculate an average deviation from the mean.

starch: a *polysaccharide* composed of long chains of *glucose* molecules.

■ Starch is actually composed of two different *polymers* of glucose: amylose (unbranched chains) and amylopectin (branched chains). The chains are usually coiled into a helix, making starch useful as a storage polysaccharide in green plants.

steroid: a type of *lipid* molecule with a number of biological functions.

■ Although steroids are classified as lipids, they do not contain *fatty acids*.

■ *e.g.* *Cholesterol* is a steroid which is an important component of *cell membranes* and is used to manufacture steroid *hormones*, such as *testosterone* and *progesterone*.

stigma: the surface at the tip of a carpel in a *flower* which is adapted to receive *pollen*.

■ Pollen is transferred to the stigma by wind or insect *vectors* during *pollination*. The pollen grains germinate on the stigma, producing *pollen tubes*, which grow down through the *style* to the ovule of the flower. This permits the fusion of male and female *gametes* during *fertilisation*.

stimulus: a change in the internal or external environment which produces a response in an organism.

■ The stimulus is detected by a *receptor* which initiates mechanisms to respond to the stimulus. In the case of nervous coordination in animals, only a sufficiently large stimulus will cause a response (see *all-or-nothing*).

■ *e.g.* A rise in the *glucose* concentration in the blood (the stimulus) will induce the secretion of *insulin* (the response) in mammals; a change in day length (the stimulus) induces flowering (the response) in plants.

stomach: part of the digestive system, located between the *oesophagus* and the *duodenum*.

■ The stomach is a muscular sac in which food is stored during the early stages of *digestion*. Cells lining the stomach produce *gastric juice*, which is mixed with food and begins to digest *proteins*. The partly digested food then passes into the small intestine, where digestion is completed and *absorption* takes place. This process is slightly modified in the digestive system of *ruminants*.

stomata: pores found in the *epidermis* of a plant.

■ Stomata are present in very large numbers, especially on the underside of leaves. They are concerned with gas exchange for *respiration* and *photosynthesis*, and also with the loss of water vapour by *transpiration*. Stomata can be opened or closed by guard cells according to the requirements of the plant, often being open in the day (to provide carbon dioxide for photosynthesis) and closed at night (to minimise water loss). The stomata of *xerophytes* (plants living in dry environments) are often adapted by being sunken into pits or surrounded by hairs to reduce water loss.

■ *TIP* Make sure that you know and understand the mechanism by which stomata open and close.

style: the stalk of a carpel in a *flower*, located between the *stigma* and the *ovary*.

■ The style is elongated in wind-pollinated plants to allow the feathery stigma to hang outside the flower and trap *pollen*.

suberin: a waxy substance consisting of long chains of *fatty acids*.

■ Suberin is found in the *endodermis* of plant roots, where it forms an impermeable *Casparian strip*. This forces water and mineral ions to pass through the *symplastic pathway* before they can enter the *xylem*. Suberin is also laid down in the *cell walls* of cork cells, providing a waterproof outer layer for many trees and shrubs.

■ *TIP* Try not to confuse suberin with *lignin*, which is another complex *polymer* found in the cell walls of certain plant tissues.

substrate: the substance acted upon by an *enzyme* in biochemical reactions.

■ Most enzymes are very specific and only one substrate will fit the *active site* to form an enzyme–substrate complex.

succession: the progressive change which occurs in a *community* of organisms over a period of time.

■ Succession starts with the initial colonisation of an area and usually progresses through several distinct stages before reaching a stable climax community. A succession which is prevented from reaching its natural climax (usually as a result of human activity) is called a *deflected succession*.

sucrose: a *disaccharide* found in plants.

■ Sucrose is composed of a *glucose* molecule linked to a fructose molecule. It is the form in which *carbohydrates* are usually transported in the *phloem* of plants.

■ *TIP* Note that sucrose is not a *reducing sugar*.

summation: a process which occurs in *synapses* due to the additive effect of a number of stimuli.

■ The arrival of an *impulse* at a synapse may not be sufficient to trigger an *action potential* in the post-synaptic cell (usually because there is not enough *neurotransmitter* released). However, two or more impulses arriving at the synapse in a short period of time may generate an action potential. This is known as temporal summation as the effects of the impulses add up over time. Similarly, an action potential may be triggered if several synapses act at the same time on the post-synaptic cell. This is known as spatial summation. The process of summation is important in controlling the movement of nerve impulses through the body.

supernatant: the liquid layer which is left on top after a suspension has been centrifuged.

surface area to volume ratio: the ratio between the surface area of an organism and its volume.

▨ If we consider a cube with a side length of 1 cm, its surface area is 6 cm² and its volume is 1 cm³. Therefore, its surface area to volume ratio (SA/V ratio) is 6. If a cube has a side length of 2 cm, its SA/V ratio falls to 3. This is a general rule: as an object increases in size, its SA/V ratio decreases. This is shown in the graph below.

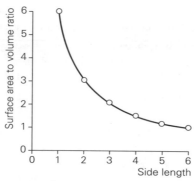

In biological terms, this rule is important because the larger the size of an organism, the smaller the SA/V ratio.

▨ *e.g.* A small one-celled organism, such as **Amoeba**, has a very large SA/V ratio. It is therefore able to gain all the oxygen it requires by *diffusion* through its body surface. A larger organism such as a badger, however, has a much smaller SA/V ratio and needs a specially adapted gas exchange surface (the *lungs*) in order to obtain sufficient oxygen.

symbiosis: a close interaction between individuals of different *species*.

▨ Symbiosis literally means 'living together' and it encompasses three different types of interaction: *mutualism, commensalism* and *parasitism*. The table below summarises the nature of these symbiotic relationships.

Symbiotic relationship	Species A	Species B
Commensalism	Gains	Unaffected
Parasitism	Gains	Loses
Mutualism	Gains	Gains

sympathetic nervous system: the part of the *autonomic nervous system* which regulates physiological functions when the body is active.

▨ The *neurotransmitters* used in the sympathetic nervous system are *acetylcholine* and *noradrenaline*. Examples of sympathetic stimulation include an increase in ventilation rate, dilation of pupils and an increase in cardiac output.

■ *TIP* Try not to confuse the sympathetic system with the *parasympathetic nervous system*, which is the part of the autonomic nervous system which regulates physiological functions when the body is at rest. The two systems generally have antagonistic effects.

sympatric speciation: the development of one or more *species*, which occurs when populations of the parent species living in the same area are prevented from interbreeding.

■ Interbreeding is prevented by various mechanisms, associated with mating, *fertilisation* and the viability of offspring. The two populations are said to be reproductively isolated from each other.

■ *TIP* A common exam question asks you to distinguish between sympatric and *allopatric speciation*, which occurs when populations of the parent species become geographically isolated from each other.

symplastic pathway: the route by which water and solutes travel through the *cytoplasm* and *plasmodesmata* of plant cells.

■ The symplastic pathway allows water and mineral ions to travel across the root cortex from the root hair cells to the *xylem*. It is the only route by which water can cross the *endodermis*.

■ *TIP* Do not confuse the symplastic pathway with the other two routes along which water can travel through plants: the *apoplastic pathway*, through the *cell walls*; and the vacuolar pathway, through the cell *vacuoles* (although this pathway still involves passing through the cytoplasm).

synapse: the junction between two nerve cells.

■ The swollen tip of the *axon* of the presynaptic *neurone* is called a synaptic knob. It contains vesicles of *neurotransmitter*, which are used to transmit the *impulse* to the postsynaptic neurone. The mechanism by which a synapse transmits an impulse from one nerve cell to the next is summarised in the flow chart below.

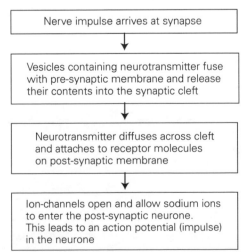

S

synergistic effect: two substances whose combined action produces a greater effect than would be expected from adding the individual effects of each substance.

■ Drugs, *hormones* and *enzymes* may all act synergistically.

■ *e.g.* *Lipases* and *proteases* work together in biological washing powders to remove stains much more effectively than would be expected from studying their individual effects.

■ *TIP* Synergistic is the opposite of *antagonistic*, which refers to substances whose combined action produces a smaller effect than would be expected from adding the individual effects of each substance.

synovial joint: a moveable joint between two bones.

■ A synovial joint is encapsulated by a tough outer layer of *collagen*. The internal ligament is lined by the synovial membrane, which secretes a fluid that lubricates the joint.

■ *e.g.* The elbow joint.

taxonomy: the study of the *classification* of living organisms.

TCA cycle: see *Krebs cycle*.

T-cell: one of a group of *lymphocytes* which play an important role in the immune response of an organism.

■ T-cells have a number of functions, including coordinating the immune response and killing cells which have been infected with pathogens. However, unlike *B-cells*, they do not produce *antibodies*.

tendon: a band of *connective tissue* that attaches *skeletal muscle* to *bone*.

■ Tendons have a high *collagen* content and so are inelastic and resistant to pulling strains. This means that the force exerted by muscular contraction is efficiently transferred to the bone.

tertiary structure: the irregular folding of a *polypeptide*.

■ The tertiary structure of a polypeptide or *protein* is determined by its *primary structure* (the sequence of *amino acids*). The shape is maintained by chemical bonds between the different amino acids, including *hydrogen bonds*, ionic bonds and bonds between sulphur-containing amino acids. The shape of a protein is closely related to its function, as seen in the *active site* of *enzymes* and the specific shape of *antibody* molecules.

testa: the protective outer covering of a *seed*.

test cross: a genetic cross used to identify the *genotype* of an individual.

■ *e.g.* In peas, the allele for a tall plant, T, is *dominant* to that for a short plant, t. A tall pea plant could therefore have one of two possible genotypes, TT or Tt. In order to find out the genotype of a particular tall plant, it is necessary to cross it with a short (tt) plant. The diagram below shows the expected results.

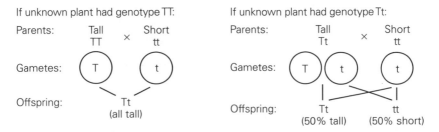

■ *TIP* See *monohybrid inheritance* for tips on how to construct genetic crosses.

testis: an *organ* that produces male *gametes*.

■ The male gametes are *sperm* and are produced by seminiferous tubules in the testes (plural) by *spermatogenesis*. The testes also produce the hormone *testosterone*.

testosterone: a sex *hormone* produced by the *testis* in males.

■ Testosterone controls the growth and development of the male sex organs and also influences *protein synthesis* and bone growth. In addition, it regulates sexual behaviour and the development of the male secondary sexual characteristics, such as a deep voice and facial hair.

thermal pollution: an increase in the temperature of the *environment* due to the release of heat from industrial processes.

■ A common form of thermal pollution is the release of warm water into rivers by power stations. This increases the temperature of the river, leading to a decrease in the concentration of dissolved oxygen in the water. This may cause a change in the nature of the *communities* in the river and result in the death of many organisms. The increase in temperature may also affect the respiration, reproduction and growth of aquatic organisms.

thermoregulation: the control of body temperature in animals.

■ In order to function normally, animals need to keep their body temperatures within certain limits. The methods used to regulate temperature vary, depending on whether the animal is an *ectotherm* or an *endotherm*. Ectotherms rely on external sources of heat and use behavioural mechanisms to regulate temperature. For example, crocodiles gain heat by basking in the sun on land, and lose heat by taking to the water. Endotherms regulate body temperature using physiological methods. For example, the diagram below summarises the regulation of body temperature in humans.

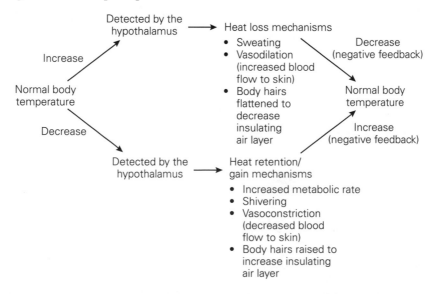

■ *TIP* Thermoregulation in endotherms is a good example of *homeostasis* (the maintenance of a constant internal environment).

thylakoid: one of a collection of flattened membrane-bound sacs found in a *chloroplast*.

■ Thylakoid membranes contain photosynthetic *pigments*, such as *chlorophyll*, and are the site of the *light-dependent reaction* of *photosynthesis*. A stack of thylakoids is known as a granum.

thymine: a *pyrimidine* nitrogenous base that is found in *DNA*.

■ In DNA, thymine always forms a bond with *adenine*, a *purine* nitrogenous base.

■ *TIP* Note that thymine is not found in *RNA*.

thyroid gland: an *endocrine* gland found in the neck of mammals.

■ The thyroid gland secretes two *hormones*: *thyroxine*, which regulates the *basal metabolic rate* of the animal; and calcitonin, which helps to control the concentration of *calcium* in the blood.

thyroid-stimulating hormone (TSH): a *hormone* secreted by the *pituitary gland* which stimulates the secretion of *thyroxine*.

■ The secretion of TSH is regulated by *thyrotrophic-releasing hormone (TRH)* from the *hypothalamus*.

thyrotrophic-releasing hormone (TRH): a *hormone* secreted by the *hypothalamus* which stimulates the secretion of *thyroid-stimulating hormone*.

■ One of the factors involved in regulating the secretion of TRH is the concentration of *thyroxine* in the blood. A low concentration of thyroxine will stimulate the secretion of TRH, ultimately leading to an increase in thyroxine concentration. This represents an example of *negative feedback*.

thyroxine: a *hormone* secreted by the *thyroid gland*.

■ Thyroxine increases the *basal metabolic rate* of cells and is involved in *thermoregulation*. It also increases cardiac output and generally makes an animal more alert. Excessive secretion of thyroxine (hyperthyroidism) causes weight loss and hyperactivity. Too little thyroxine (hypothyroidism) causes retarded physical and mental development in young animals, and lethargy and obesity in older animals. It is important, therefore, to regulate the secretion of thyroxine. This is achieved by a *negative feedback* mechanism, as shown in the diagram below.

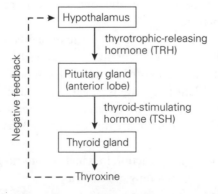

tissue: a collection of similar cells which are specialised for a particular function.

▓ Some tissues consist of several different types of cell. For example, blood consists of *red blood cells, white blood cells* and *platelets*. Different tissues are grouped together to form *organs*, such as a *flower* or a *kidney*.

▓ *e.g.* *Xylem* tissue in plants and *muscle* tissue in animals.

▓ *TIP* It is useful to know the different levels of organisation in living organisms, starting at the simplest (smallest) level: *organelle — cell — tissue — organ —* organ system — organism.

tissue fluid: the fluid which surrounds the cells in an animal.

▓ Tissue fluid is formed when blood *plasma* is forced out of *capillaries* by *ultra-filtration*. High pressure at the arterial end of the capillaries forces the plasma out of the capillaries and into the intercellular spaces. Tissue fluid has the same composition as plasma, except that it lacks *plasma proteins* which are too large to leave the capillaries. It is the medium from which cells take up nutrients and respiratory gases, and into which they release waste products. At the venous end of the capillaries (where pressure is low), the tissue fluid is taken back into the blood by *osmosis*. Any excess fluid around the cells is taken up by *lymph* vessels.

trachea (mammal): a tube which links the mouth and nose to the *lungs* in air-breathing vertebrates.

▓ The trachea is also known as the windpipe and runs from the larynx (voice-box) to the bronchi. It is supported by rings of *cartilage* which prevent it collapsing during ventilation. The inner surface is lined with ciliated *epithelium* which secretes *mucus*. Dust and microorganisms are trapped in the mucus which is moved up to the throat by the action of *cilia*. It is then swallowed and broken down by the digestive system.

tracheae (insect): tubes which form part of the gas exchange system in an insect.

▓ The tracheae are tubes containing rings of *chitin* which lead from the *spiracles* in the *exoskeleton* of an insect and run to all areas of the body. They branch into microscopic tubules called tracheoles, providing a large surface area for gas exchange. Oxygen diffuses into the tracheoles and dissolves in tissue fluid at the end of the tubules before being absorbed by the cells. When the insect is active, pumping movements of the abdominal muscles cause air to be drawn in and out of the tracheae.

tracheid: a type of cell found in the *xylem* tissue of plants.

▓ Tracheid cells are elongated and contain large quantities of *lignin* in their *cell walls*. Their end-walls are perforated by a number of pores which allow water to move through the plant from one tracheid to the next.

▓ *TIP* Make sure that you know the difference between tracheids and *vessels*, which are also found in xylem but are not separate cells as they have lost their end-walls.

transamination: a biochemical reaction in which an amino group is transferred from one *amino acid* to form a new amino acid.

■ Transamination takes place in the *liver* of mammals. The process is summarised in the diagram below.

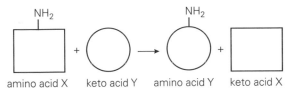

transcription: the copying of the *genetic code* from *DNA* to *messenger RNA (mRNA)* during *protein synthesis*.

■ During transcription, part of a DNA molecule (a *gene*) unwinds and one strand acts as a template for the synthesis of a molecule of mRNA (catalysed by *RNA polymerase*). For example, the DNA sequence AGCTATCGA would be transcribed into the mRNA sequence UCGAUAGCU. The mRNA strand then leaves the *nucleus* and becomes attached to a *ribosome* in the *cytoplasm* of the cell.

■ *TIP* Try not to confuse transcription with *translation*, which is the process by which the sequence of bases in mRNA is translated into a sequence of *amino acids* during protein synthesis.

transect: a technique used in ecology to sample the organisms across an area.

■ Transects sample every organism in contact with a line drawn across an area. They can be used to investigate zonation within a *community*, such as the distribution of organisms on a rocky sea shore.

transfer RNA (tRNA): a type of *RNA* which carries *amino acids* to a *ribosome* during *protein synthesis*.

■ A molecule of tRNA consists of a single *polynucleotide* chain twisted into a hairpin shape. Each molecule can carry a specific amino acid and contains a triplet of bases known as an *anticodon*. This enables tRNA to line up opposite its complementary *codon* on *messenger RNA (mRNA)* during *translation*. Transfer RNA is therefore important in assembling amino acids in the correct order during the synthesis of a protein.

translation: the process by which the sequence of bases in *messenger RNA (mRNA)* is translated into a sequence of *amino acids* during *protein synthesis*.

■ Translation takes place on the *ribosomes* in a cell. Each *codon* on an *mRNA* molecule pairs with an *anticodon* on a transfer RNA (tRNA) molecule, which carries a specific amino acid. As the ribosome moves along the mRNA molecule, the amino acids are joined by *condensation* reactions to form a *polypeptide*. The sequence of amino acids in the polypeptide (its *primary structure*) is determined by the sequence of bases in mRNA, which itself is determined by the sequence of bases in *DNA* (the *gene*). Therefore, during protein synthesis, one gene codes for the production of one polypeptide.

■ *TIP* Try not to confuse translation with *transcription*, which is the copying of the *genetic code* from DNA to mRNA during protein synthesis.

translocation: the transport of substances from one part of a plant to another.

■ Water and mineral ions are transported through the plant in the *xylem* according to the *cohesion–tension theory*. Organic molecules, such as *sucrose* and *amino acids*, are transported from a site of production (e.g. the photosynthetic cells in a leaf) to other parts of the plant via the *phloem*. In other words, they move from a *source* to a *sink*. Although most biologists believe that substances move through the phloem by a system of mass flow, the precise mechanisms are unknown.

transpiration: the loss of water vapour from a plant.

■ Transpiration occurs mainly through the *stomata* on leaves, but also to a lesser extent through the *lenticels* on woody stems. This loss of water vapour creates a *water potential* gradient which draws water up the *xylem* according to the *cohesion–tension theory*. The rate of transpiration will be increased by: increasing the temperature, light intensity or wind speed; or decreasing the relative humidity around the plant.

TRH: see *thyrotrophic-releasing hormone*.

tricarboxylic acid (TCA) cycle: see *Krebs cycle*.

tricuspid valve: the valve between the right *atrium* and the right *ventricle* in the heart of a mammal.

■ *TIP* Do not confuse the tricuspid valve with the *bicuspid valve*, which is found between the left atrium and the left ventricle. It may help to think about the t**R**icuspid valve being on the **R**ight. Both valves are also known as atrioventricular valves.

triglyceride: a *lipid* molecule consisting of three *fatty acids* linked to a molecule of glycerol.

■ The general structure of a triglyceride is shown in the diagram below.

$$
\begin{array}{c}
H \\
| \\
H-C-O-FA_1 \\
| \\
H-C-O-FA_2 \\
| \\
H-C-O-FA_3 \\
| \\
H
\end{array}
$$

FA = Fatty acid

Ester bond

The physical and chemical properties of the triglyceride depend upon the nature of its fatty acids. Triglycerides are the major constituents of fats and oils and provide a concentrated energy store in living organisms.

triploblastic: any animal with a body composed of three layers of cells.

■ The three layers of cells are the ectoderm, mesoderm and endoderm. The ectoderm is the outer layer and forms the *epidermis* and nervous tissue of the animal. The endoderm is the inner layer, forming the lining of the digestive system. The mesoderm lies between the other two layers and forms *connective tissue* and all the other organs. All animals are triploblastic except those in the *phylum Cnidaria*, which are *diploblastic*. Many triploblastic animals possess a *coelom*, formed from the mesoderm, which allows the development of specialised organs for both locomotion and *digestion*.

tRNA: see *transfer RNA*.

trophic level: the position occupied by an organism in a *food chain*.

■ The first trophic level is occupied by *producers*, such as green plants. The second trophic level contains *primary consumers* (herbivores) which feed on the producers. The third trophic level contains secondary consumers (carnivores) which feed on the primary consumers, and so on. There are rarely more than five trophic levels in a food chain as energy is lost during the transfer from one level to the next (see *pyramids of numbers/biomass/energy*). The diagram below shows the trophic levels of organisms making up a typical food chain.

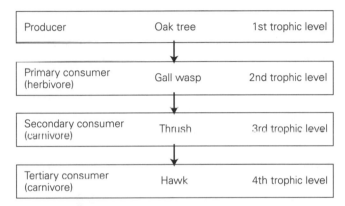

■ *TIP* Note that the ultimate trophic level is that of the *decomposers*, which release *inorganic* nutrients for recycling via the producers. One problem with using trophic levels is that many animals feed at more than one trophic level, making it difficult to define feeding relationships accurately.

tropism: the directional *growth* of part of a plant in response to an external *stimulus*.

■ *e.g.* Most stems are positively phototropic, because they grow towards light, and negatively geotropic, because they grow away from gravity.

trypsin: a protein-digesting *enzyme* produced by the *pancreas*.

■ Trypsin is secreted in pancreatic juice in an inactive form, trypsinogen, to avoid self-digestion by the secretory cells. It is activated to trypsin by the *enzyme* enterokinase, which is produced by cells lining the small intestine. Trypsin is an *endopeptidase*, hydrolysing the *peptide* bonds within proteins and converting them into shorter *polypeptides*.

TSH: see *thyroid-stimulating hormone*.

turgid cell: a cell which contains the maximum amount of water.

■ The *water potential* of a fully turgid cell is zero (because the *solute potential* and the *pressure potential* are equal). Therefore, no more water can enter the cell; it can only move out. Turgidity is important in providing support for plants and for the opening and closing of *stomata*. A loss of turgor will lead to *wilting*.

ultrafiltration: filtration which occurs on a molecular scale.

■ *e.g.* High blood pressure in the *capillaries* of the *glomerulus* forces fluid through the basement membrane into the *Bowman's capsule* of a kidney *nephron*. This fluid is known as *glomerular filtrate* and has the same composition as blood *plasma*, except that it does not contain *plasma proteins* (which are too large to pass through the basement membrane of the capillaries). *Tissue fluid* is also formed by ultrafiltration. High blood pressure at the arterial end of capillaries forces fluid (plasma minus the large proteins) through the capillary walls into the surrounding tissues.

ultrastructure: the structure of cells and *organelles* as seen using an electron microscope.

■ Electron microscopes have a much greater resolving power than light microscopes and are able to distinguish the fine detail of cell structure. The detailed structure of many organelles, such as *ribosomes*, is not visible under a light microscope.

unsaturated fatty acid: a *fatty acid* which has less than the maximum number of hydrogen atoms attached to its carbon atoms.

■ Unsaturated fatty acids have at least one double bond between their carbon atoms.

■ *e.g.* Oleic acid, $CH_3(CH_2)_7C=C(CH_2)_7COOH$.

■ *TIP* Try not to confuse unsaturated fatty acids with *saturated fatty acids*, which have no double bonds between their carbon atoms.

uracil: a *pyrimidine* nitrogenous base that is found in *RNA*.

■ During *protein synthesis*, uracil will form a bond with *adenine*, a *purine* nitrogenous base.

■ *TIP* Note that uracil is not found in *DNA*.

urea: the main nitrogenous waste product of mammals.

■ Excess *amino acids* in the body are deaminated by the *liver*, producing ammonia as a toxic waste product. The ammonia is then converted to less-toxic urea via the *ornithine cycle*. Urea is transported in the blood to the *kidneys*, where it is excreted in the form of *urine*.

■ *TIP* Students often confuse *urine*, *urea* and *glomerular filtrate*. Make sure that you know and understand the differences between these terms.

urine: an aqueous solution produced by the *kidneys* in mammals, that contains *urea*, inorganic salts and water.

■ Urine passes from the kidneys along the ureters to the bladder, where it is stored before being removed from the body via the urethra. The actual composition of urine is variable and can be influenced by a number of factors, including climate and diet.

■ *TIP* Students often confuse urine, urea and *glomerular filtrate*. Make sure that you know and understand the differences between these terms.

uterus: the organ in which the *embryo* develops in female mammals.

■ The uterus is also known as the womb and is connected to the oviducts and the vagina. The lining of the uterus is called the *endometrium* and it shows cyclical changes associated with egg production, known as the *menstrual cycle*. During pregnancy, the endometrium helps to nourish the young embryo before the *placenta* is established. The outer wall of the uterus is thick and muscular, and its contractions are responsible for the process of birth.

vacuole: a membrane-bound space within the *cytoplasm* of a cell.

■ A vacuole may contain air, water, cell sap (a solution of sugars, *amino acids*, inorganic salts and waste products), or food particles. Plant cells usually have one large vacuole which is bounded by the tonoplast (vacuolar membrane) and contains cell sap. Animal cells do not usually have vacuoles. Certain *protoctists*, such as **Amoeba**, have a *contractile vacuole*, which is specialised for *osmoregulation*.

variation: the phenotypic differences between individuals within a *species*.

■ Variation may be due to the effects of the *genotype* and/or the *environment*. Genetic variation is due to: independent assortment of *homologous chromosomes* and *crossing-over* during *meiosis*; random fertilisation between male and female *gametes*; and *mutation*. *Selection* acts on these genetic variations, determining which characteristics survive from one generation to the next. Phenotypic characteristics show two forms of variation: *discontinuous variation* (clearly defined differences within a population) and *continuous variation* (a complete range of forms can be seen within a population).

vascular system: a system of vessels for the transportation of substances from one part of an organism to another.

■ Vascular systems are associated with *mass transport*.

■ *e.g.* The blood system of a mammal and the vascular tissue (*xylem* and *phloem*) in plants.

vasoconstriction: a decrease in the internal diameter of an *arteriole*.

■ Vasoconstriction is caused by contraction of *smooth muscle* in the wall of an arteriole, reducing the size of the *lumen*. This process is regulated by the *autonomic nervous system*. Vasoconstriction is important in *thermoregulation*, as it reduces blood flow to the skin surface and so conserves heat. Vasoconstriction may also be used to increase blood pressure.

■ *TIP* Note that vasoconstriction does not happen in *capillaries*, as these vessels do not contain smooth muscle.

vasodilation: an increase in the internal diameter of an *arteriole*.

■ Vasodilation is caused by relaxation of *smooth muscle* in the wall of an arteriole, increasing the size of the *lumen*. This process is regulated by the *autonomic* nervous system. Vasodilation is important in *thermoregulation*, as it increases blood flow to the skin surface and so promotes heat loss. Vasodilation may also be used to decrease blood pressure.

■ *TIP* Note that vasodilation does not happen in *capillaries*, as these vessels do not contain smooth muscle.

vector: any organism which acts as a carrier.

■ The term vector generally has two meanings in biology.
- In *gene technology*, a vector is anything that can be used to insert foreign *DNA* into the genetic material of a host cell, usually a *bacterium*. Examples of such vectors include *plasmids* and bacteriophages.
- In pathology, a vector is an animal that passively transmits disease-causing microorganisms from one organism to another. Examples of such vectors include aphids, which spread viral diseases in plants, and mosquitoes, which transmit malarial *parasites* between humans.

vein (animal): a vessel that carries blood towards the *heart*.

■ All veins carry deoxygenated blood, except the *pulmonary vein*, which transports blood from the *lungs* to the heart. The general structure of a vein is shown in the diagram below.

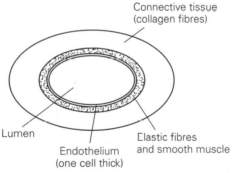

Venous blood is at low pressure, so the walls of the veins contain much less muscle and elastic tissue than those of *arteries*. Contractions of the skeletal muscles around the veins squeeze blood along the vessels and *semilunar valves* help to prevent back flow. As a result, the flow of blood back to the heart is maintained.

vein (plant): part of the *vascular system* in the leaves of plants.

■ Veins help to support the leaf and contain the vascular tissue. This consists of the *xylem*, which supplies the leaf with water and mineral ions, and the *phloem*, which transports the products of *photosynthesis* to the rest of the plant.

vena cava: the main *vein* of the body.

■ The vena cava carries deoxygenated blood from all parts of the body back to the right *atrium* of the *heart*. All other veins (except the *pulmonary vein*) feed into the vena cava.

ventilation: the mechanism by which respiratory gases are delivered to and from a gas exchange surface.

■ Gas exchange surfaces in animals take up oxygen and release carbon dioxide. In order that these processes occur efficiently, the air or water in contact with the gas exchange surface must be changed regularly. This maintains a *diffusion* gradient for the respiratory gases. Different animals have different ventilation mechanisms.

- Insects use abdominal muscles to pump air in and out of the *spiracles* and *tracheae*.
- Fish pump water over the surface of the *gills*.
- Mammals use muscles in the diaphragm and between the ribs to pump air in and out of the *lungs*.

In mammals, the rate of ventilation is carefully controlled to ensure that body cells have an adequate supply of oxygen for *aerobic respiration*. The diagram below summarises the processes involved in regulating the ventilation rate.

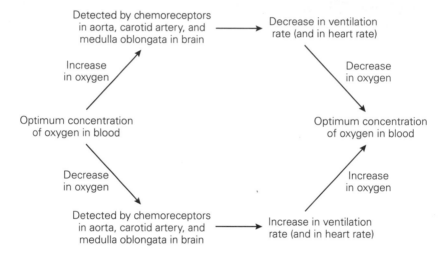

ventricle: a chamber of the *heart* that pumps blood to the organs of the body.

■ In mammals, the right ventricle pumps deoxygenated blood to the lungs via the *pulmonary artery*, and the left ventricle pumps oxygenated blood to the other organs of the body via the *aorta*.

■ *TIP* Try not to confuse ventricle with *atrium*, which is a chamber of the heart that receives blood returning from the organs of the body.

venule: a vessel that receives blood from *capillaries* and carries it to a *vein*.

■ Venules have a smaller diameter than veins, but have the same general structure.

vesicle: a small membrane-bound sac found in the *cytoplasm* of a cell.

■ Vesicles are used to carry substances which have been taken into the cell by *endocytosis*, or which are being secreted from the cell by *exocytosis*. The *Golgi apparatus* is responsible for processing and packaging substances produced by the cell and it is therefore associated with a number of vesicles. Apart from the fact that they are smaller, vesicles are essentially the same as *vacuoles*.

■ **e.g.** At a *synapse*, the synaptic knob contains numerous vesicles of *neurotransmitter*. When stimulated, these vesicles release their contents into the synaptic cleft, propagating the nerve *impulse*.

vessel: a type of cell found in the *xylem* tissue of plants.

■ Vessel cells are elongated and contain large quantities of *lignin* in their *cell walls*. Their end-walls are broken down during development to provide connections with other vessel cells above and below (rather like a drainpipe). These allow efficient movement of water through the plant.

■ *TIP* Make sure that you know the difference between vessels and *tracheids*, which are also found in xylem but are separate cells with pores in their end-walls.

villus: a finger-shaped projection of the lining of the small intestine.

■ The general structure of a villus is shown in the diagram below.

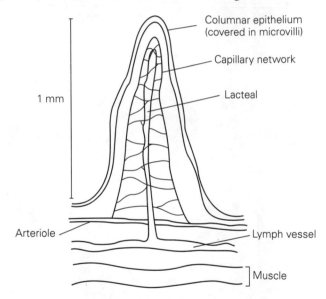

Villi (plural) occur in very large numbers throughout the small intestine. They are covered with columnar *epithelium*, which itself has microvilli (microscopic projections of the epithelial cell surface membranes), so there is a very large surface area for *absorption* of the products of *digestion*. Each villus has an extensive *capillary* network to transport the absorbed products to the *liver* (except *fatty acids* and glycerol, which enter the lymphatic system via a lacteal).

■ *TIP* Note that villi and microvilli are different structures and should not be confused.

virus: a *microorganism* which can only reproduce inside a living cell.

■ Viruses range in size from 20 to 400 nm in diameter. They consist of a core of *nucleic acid* (*DNA* or *RNA*) surrounded by a *capsid* (protein coat). Outside its host cell, a virus is completely inert and carries out none of the processes associated with living organisms. Inside its host cell, however, the virus can

replicate and the new viruses are released when the cell breaks down. Viruses cause a wide range of human diseases, including rabies, influenza and AIDS. They can also cause a number of plant diseases.

vitamin: an *organic* molecule required by living organisms in small amounts in order to maintain normal health.

■ Vitamins act mainly as *coenzymes* in metabolic processes. Most vitamins cannot be synthesised by the body and therefore must be ingested in the diet.

■ *e.g.* Vitamin C (ascorbic acid) is a water-soluble vitamin which is found in relatively large quantities in citrus fruits and green vegetables. It promotes protein metabolism and the synthesis of *collagen*, thus helping in wound healing. A deficiency of vitamin C leads to scurvy, characterised by impaired wound healing and bleeding of the gums.

water potential: a measure of the ability of a solution to give out water.

■ Water potential (ψ) can be expressed as the sum of the *solute potential* (ψ_S) and the *pressure potential* (ψ_P):

$$\psi = \psi_S + \psi_P$$

The water potential of pure water is zero and all other solutions have a negative water potential. The greater the solute concentration (or the lower the relative concentration of water), the more negative is the value for water potential. Water will move from areas of high (less negative) water potential to areas of low (more negative) water potential, i.e. down a water potential gradient.

■ *e.g.* In the two cells shown below, water will move from B to A by *osmosis* because B has a higher (less negative) water potential.

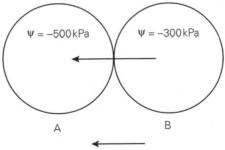

$\psi = -500\,\text{kPa}$ $\psi = -300\,\text{kPa}$

A   B

Direction of net water movement

■ *TIP* If you get confused about the negative numbers, try thinking about temperature. A temperature of $-2\,°C$ is clearly higher (less negative or closer to zero) than one of $-10\,°C$.

white blood cell: a nucleated cell found in the blood.

■ There are several different types of white blood cell (leucocyte), but they are all involved in coordinating the immune response. For example, *lymphocytes* produce *antibodies,* and neutrophils and monocytes ingest *bacteria* by *phagocytosis.*

wilting: the condition that occurs in plants when more water is lost by *transpiration* than is absorbed from the soil.

Net water loss causes the plant cells to lose their *turgidity* and the plant will droop. Plants can normally recover from wilting if water is added to the soil within a reasonable period of time. In certain plants wilting is an important mechanism to avoid overheating (when the leaves droop they are taken out of direct sunlight). When the sun sets, these plants reduce their rate of transpiration and the cells regain their turgor.

X chromosome: one of the *sex chromosomes*.

■ Sex (*male* or *female*) is often determined by *genes* carried on the sex chromosomes. Female mammals have two X chromosomes and are known as the homogametic sex. Male mammals have one X and one *Y chromosome* and are known as the heterogametic sex.

xerophyte: a plant which is adapted for growing in dry environments.

■ Xerophytes usually have a number of adaptations to reduce water loss by *transpiration* and to maximise water uptake from the soil. They have:

- leaves which are reduced to spines to minimise the surface area available for transpiration (in plants with photosynthetic stems), e.g. gorse
- rolled leaves with sunken *stomata* and a thick cuticle, e.g. marram grass
- root systems which cover a very wide area or penetrate deep into the soil, together with water storage in tissues, e.g. cacti

xylem: the plant vascular tissue concerned with the transport of water and inorganic salts.

■ Xylem is composed of *tracheids* and *vessels*. Water generally moves up the xylem from the roots to the stem and leaves, as described by the *cohesion–tension theory*. The xylem also has an important supporting role in the plant.

Y chromosome: one of the *sex chromosomes*.

■ Sex (*male* or *female*) is often determined by *genes* carried on the sex chromosomes. Female mammals have two *X chromosomes* and are known as the homogametic sex. Male mammals have one X and one Y chromosome and are known as the heterogametic sex.

zygote: a *diploid* cell that results from the fusion of two *gametes* during *sexual reproduction*.